Life Concepts from Aristotle to Darwin

Lucas John Mix

Life Concepts from Aristotle to Darwin

On Vegetable Souls

Lucas John Mix
Department of Organismic and
   Evolutionary Biology
Harvard University
Cambridge, MA, USA

ISBN 978-3-030-07139-4   ISBN 978-3-319-96047-0   (eBook)
https://doi.org/10.1007/978-3-319-96047-0

Cover image: GeorgePeters/Getty Images
Cover design: Akihiro Nakayama

This Palgrave Macmillan imprint is published by the registered company Springer Nature
Switzerland AG
The registered company address is: Gewerbestrasse 11, 6330 Cham, Switzerland

Preface

When I first encountered the vegetable soul, I thought it an odd curiosity, a strange turn of phrase. As I investigated, I began to realize it had an important part to play in the history of "life" as we know it. Vegetable souls oriented biology for two millennia through the influence of Plato, Aristotle, Lucretius, Pliny, Origen, Plotinus, Ibn Sînâ, Maimonides, and Aquinas. They influenced the shift to modern biology through Gassendi, Leibniz, and Hegel. And yet, I could find no systematic account, no story of vegetable souls. The very term seemed laden with contradiction and was unknown to many historians and philosophers.

I set out to write the book I wanted to read: a history of theories of life prior to Darwin, in particular a history of vegetable souls.

As we move forward in biology, exploring the origins of life on Earth, the possibility of life elsewhere, and the very boundaries of individuality (genetic, microbial, and neurological), it seems worthwhile to review where we have been. Some of the connections we wish to make, between physics and biology, biology and humanity, have been proposed before. Past successes (and failures) can help us evaluate which concepts work for us.

Such a project would not have been possible without generous support from multiple sources. David Haig, John Wakeley, and the Harvard Department of Organismic and Evolutionary Biology graciously hosted me and granted me access to the amazing Ernst Mayr Library. Parts of this project were made possible through the support of a grant from the John Templeton Foundation. (The opinions expressed in this publication are those of the author and do not necessarily reflect the views of the

John Templeton Foundation.) I was also greatly aided by a year at the Center of Theological Inquiry, where I was a resident for the 2015–2016 inquiry into the "Societal Implications of Astrobiology." During that year, I was able to discuss life-concepts in depth with theologians and philosophers. The number of people who inspired, informed, and corrected me is too great to name.

I owe specific thanks to Andrew Accornero, Andrew Berry, Bill Brown, Luis Campos, David Haig, and Matt Quarterman for reading chapters and providing comments. Thanks are due as well to my family and colleagues who put up with five years of postponing other projects to pursue this topic. I am deeply grateful.

Cambridge, USA   Lucas John Mix

CONTENTS

CHAPTER 1

Vegetable Souls?

The term "vegetable soul" troubles the ear. The two words have such different contexts that joining them strikes us as either funny or nonsensical. It conjures images of tortured carrots and zombie broccoli. Surprisingly, vegetable souls were one of the most popular ways of describing life prior to the Renaissance. The Aristotelian idea of souls as the efficient, formal, and final cause of life dominated discussions of plant life in Europe and adjacent regions from classical antiquity through the High Middle Ages, roughly 2000 years. And yet the idea has disappeared from our vocabulary. A great divide has opened up between the way we speak of rocks and organisms and the way we speak of persons. Vegetables belong to one discourse; souls to another. As scientists and philosophers try to reconnect the two ways of thinking, it is worth looking back at the history of vegetable life and the many attempts to find one theory of life that spans vegetable, animal, and rational domains.

We have strong intuitions about what it means to be alive, but those intuitions are constantly challenged by new discoveries. Microbiologists have uncovered vast realms of tiny organisms living in extreme environments, and even within humans. Astrobiologists look to the stars and ask whether still stranger forms of life might be out there (Hays et al. 2015). Both have provoked us to revisit systematic questions about what it means to be alive. Meanwhile, ethical debates, from genetic modification to human cell cultures, from embryos to organ transplants, stretch the meaning and boundaries of life. They have a profound impact on how we understand biological fitness as well as ethical value. The history of

© The Author(s) 2018
L. J. Mix, *Life Concepts from Aristotle to Darwin*,
https://doi.org/10.1007/978-3-319-96047-0_1

vegetable souls provides a broader perspective, more options to consider, and prospects for future research.

This book looks at the history of souls, purpose, and agency as they have been applied to the non-human, non-animal living world. Such a vast swath of reality now includes plants, fungi, protists, bacteria, and archaebacteria (and perhaps viruses). All of these were traditionally lumped together as "vegetables" and considered the lowest form of life. Most thinkers believed that vegetables have less freedom, dignity, and drive than animals, but more than mere minerals. Life was seen as a continuum, stretching from the lowest lichen to the smartest man, and perhaps beyond.[1]

Vegetable and animal souls were mortal and natural by most understandings of those words and fell into a middle category between the non-living environment and human life. They were the province of natural philosophers, the precursors of modern scientists, as well as theologians. Christians, Muslims, and Jews claimed—and still claim—that humans have a special kind of soul, an eternal soul. That claim, however, once came in the context of a larger theory of life. It came as one aspect of a soul-theory that spoke to the life of plants and animals. We were the "rational animal," but no less animal for that. We lived among a variety of living, ensouled things.

In the Renaissance and Enlightenment, theories of life changed dramatically. Cartesian dualism pushed the human soul out of the natural world, along with will, reason, and agency. Vegetable and animal souls, meanwhile, disappeared. They were neglected in the humanities and intentionally ejected from the natural sciences. Empiricists considered them unknowable, along with formal and final causes. The move proved tremendously successful in the development of modern biology, but it prohibited any common understanding that would bridge human and non-human life.

The medieval understanding of souls—as subsistent, immaterial agents—will never fit with modern science. Aristotle's picture, however,

---

[1] From a twenty-first century perspective, I can object to both ends of this scale. Lichens represent a sophisticated symbiosis of fungi, algae, and cyanobacteria. No matter how we define "simple," many simpler organisms exist. Likewise, I would object to the superiority of men over women. I might even ask how we are to define humans as superior to other multicellular organisms. I will return to these questions in part IV. As to animate life with more dignity than humans, the traditional contenders were stars, planets, and angels.

proves more promising. His idea of the vegetable soul—an active, chemical process whose cause, identity, and purpose is self-perpetuation—fits surprisingly well with evolutionary theory. By reviving Aristotle's vegetable souls as dynamic processes of nutrition and reproduction, we can come to a better understanding of biological individuality in both animals and plants. We can also see how dramatically it differs from the subjective individuality so central to the modern humanities. The vegetable soul can help us recapture and reimagine the continuity of humans with other life.

The term itself may not be recoverable, but we need ways of talking about intermediate levels of purpose and agency. This volume takes a closer look at how vegetable life has been understood through the centuries to prepare the way for a new understanding of life.

Goals for the Book

This book started as an exercise in biology. Scientists require a theory of life when asking questions at the boundaries of life. For the past 20 years, I have been involved in research into the origins of life on Earth and the extent of life in the universe. At first glance, the questions seem large and abstract, but they can quickly become concrete. In 1984, scientists recovered a meteorite from Antarctica. Chemical analysis clearly demonstrated that the rock originated on Mars, but a controversy arose in 1996. Scientists at Johnson Space Center discovered carbon globules within the meteorite. They have the shape, if not the size, of fossilized bacteria and they contain carbon and magnetite structures consistent with known life (McKay et al. 1996). This raised a question for researchers. Can we distinguish between materials generated by life and those that arise abiologically? Debate raged over whether this was proof of Martian life. Eventually, the scientific community decided that the evidence was insufficient to the claim, but it established the need for serious, cross-disciplinary research into the boundaries of life. The *Curiosity* rover is currently exploring Mars, looking for signs of life, and more rovers are planned for the future. The astrobiology community arose around this research program: looking for life. Theories of life have concrete, practical consequences as we explore Mars, Europa, and points beyond.

Researching life-concepts in science, I began to notice a trend. Scientists (and philosophers) propose definitions of life, unwittingly echoing earlier proposals. In the last century, this has frequently been accompanied by promises that current knowledge would reveal a new and definitive theory of life along with the synthesis of life in the laboratory. Despite great advances in biochemistry, there is still no consensus on a definition, though self-replication and self-regulation stand out as the most promising prospects. Nor has life been synthesized. The further back I looked, the more I saw repetitions of the same themes without definitive arguments.

A few sources from the early twentieth century referred to vegetable souls as natural, material, drivers of life. At first, it seemed to be a rare oddity, something rejected by the founders of modern biology. As I looked deeper, however, I began to see references to Aristotle's vegetable souls appearing regularly, even in the work of Ernst Mayr and other proponents of the modern synthesis. Renewed interest and reinterpretation led to a revival of Aristotelian thinking in the 1970s. Modern biology was not as alien to vegetable souls as I had thought (Mix 2016).

Over the past 20 years, many authors have tackled the history of souls, emphasizing the development of concepts of self, mind, and subjectivity from pre-Socratic philosophy to cognitive neuroscience. Excellent examples include Martin and Barresi (2006) and Goetz and Taliaferro (2011). None of these works, however, give more than a passing mention to vegetable souls and the activities associated with them: nutrition and reproduction. In the same period, scientists have narrowed the gap between plants and humans, a gap believed insurmountable only 50 years ago. Increasingly, we can connect human behavior with biochemical, neurological, and genetic processes. Plant behavior looks far more complex than anyone imagined. Philosophers have taken this as an opportunity to re-evaluate how we think of humans, plants, and their relationship (e.g., Marder 2014; Nealon 2015; Marder and Irigaray 2016). The time seems ripe for an investigation of how humans and plants have been linked historically.

The first three sections of this book follow the long and complicated history of vegetable souls and the vegetable life-concept. History always requires a balance between historical detail, compelling narrative, and engagement with current scholarship. Because the term is so unfamiliar, I have leaned toward historical detail, providing a broad chronological coverage focusing on primary sources. It is my sincere hope that others will use this as a jumping off point for more detailed analysis. First, the term must be brought back into common awareness, if only as an influential aspect of historical thought. Biology has existed for millennia, both as a unified subject matter and as a way of thinking. It studies living beings, including plants, animals, and humans. For 2000 years, it invoked souls as their motivator and organizer. Part I covers the birth of vegetable souls in Ancient Greek thinking, focusing on Plato and Aristotle. Part II explores the development of vegetable souls through the Middle Ages. Influenced by Jewish, Christian, and Muslim theology, souls took on a strongly Platonic character. Part III explains the demise of vegetable souls during the Renaissance and Enlightenment, as Platonism was replaced by other approaches.

In Part IV, I ask whether vegetable souls should be resurrected for modern biology. Though I do not think they can be revived in full, aspects of Aristotelian soul thinking deserve renewed attention. Stripped of Platonic accretions, vegetable souls may still prove useful. Life looks a great deal like an active, chemical processes whose cause, identity, and purpose is self-perpetuation. The sticky parts of this—identity as formal cause and purpose as final cause—become empirically tractable in light of evolution. A nominalist approach to the formal cause allows us to see multiple and non-exclusive, biological individuals all around us. Evolution by natural selection provides a way to understand final causes that is both scientifically rigorous and biologically useful. We may be forced to decide, with Aristotle and Darwin, that individuality and purpose mean something slightly different than we thought. And yet, they are still meaningful.

Recurrent Themes

Throughout the book, I track four interrelated themes: composition, agency, individuality, and purpose. They are related to Aristotle's four causes and, thus, to his definition of a soul: an identity of efficient, formal, and final causes. Life-concepts critically impact the categories we use and the accounts we consider acceptable when describing the world. Divisions between mineral, vegetable, animal, and rational beings permeate common usage among philosophers, among biologists, and even in public. Two examples bring this home. Descriptions of living things invoke purpose (e.g., the heart is *for* pumping blood). Descriptions of human behavior invoke agency (e.g., Maurice was responsible for the crash). In both cases, adequate descriptions of in-group subjects require a unique type of explanation; acceptable descriptions outside the group prohibit it. Recent developments in biology and philosophy have called traditional divisions into question and it is worth looking at how we draw these boundaries.

I have attempted to map the shifting borders, demonstrating both the origins and contingency of current conceptions. Historical thinkers did not divide the world in familiar ways and their life-concepts cannot be understood if we attempt to fit them into modern categories. Notably, historical concepts of souls (up to and including Christian treatment of human souls) were not univocally immaterial, intentional, atomistic, or subsistent. Concepts such as imagination, sensation, and agency have been smeared across the vegetable-animal-rational spectrum. This makes terminology difficult. I beg your indulgence in allowing me some linguistic leeway. I have sought to balance descriptive consistency with historical accuracy. The next few subsections set forth my use of philosophical terms. Non-technical readers may wish to skip ahead to the section on Plato and Aristotle.

Composition

What are living things made of? The term *substance* refers to that which grounds or stands under other things. A generic definition suggests that substance describes the fundamental entities in any philosophical system. They attempt to capture the basic character underlying phenomena, that which persists through change. More specific definitions of substance require ontological commitments about the types of things that can exist and their essential features. Such specific definitions might produce a system where other, non-substantial basic entities exist (e.g., events, experiences, forces) or even replace the narrowly defined "substance." Aristotle, for example, held to forms as substances, while David Hume, maintained that persistent substances are both unintelligible and non-existent. I use the term substance in the generic sense. Are living things substances? Are they indivisible and distinct? If not, of what substances are they composed?

Changing ontologies insure that many words, now considered synonymous, had distinct meanings at distinct times. For the sake of consistency, I use several words with a restricted meaning. *Material* and *corporeal* refer to bodies (i.e., spatial extension). They exist here but not there, now but not then. They need not have distinct limits. For example, a cloud is physically located, even though it is not discretely bounded. I use *physical* and *natural* to refer to things which follow the same basic rules as particular, tangible stuff.[2] Alternatives have included material but non-physical souls which permeate the body and immaterial souls that only interact with matter miraculously. An entity may still be physical and natural if it is further constrained by other rules (as an Aristotelian final cause acting through efficient causes). It simply cannot break the general rules of physics. Three examples clarify the difference. A ghost would be corporeal (it has a visible shape), but not physical (it passes through walls). An electron is incorporeal (it has no volume), yet physical (it obeys physical laws—even when it passes through walls). The soul seeds of Lucretius are both corporeal (they have volume) and physical (they operate by natural laws), but not physical in the modern sense of the word (they have intrinsic motion).

Causal Processes

Living things participate in causal processes in a special way. At least they seem to. Whatever position we take on the metaphysics of causation, we must recognize a unique language for accounts of life and living. We speak normatively and teleologically about even the simplest organisms. We can speak of an *Escherichia coli* cell having fitness and function or, less cautiously, having success and purpose.

Debate persists about the best way to interpret this language. On one side, strict physical reductionists claim that all causes can be fully understood as matter in motion. Thomas Hobbes, T. H. Huxley, and Jacques Monod stand out as clear examples. All living things, including humans, are simply and exclusively collections of matter moving as part of causal chains that start before and beyond them. Life-specific causation is an illusion. At best, we might claim that the living being represents a specially indexed link in the chain. We attribute significance; no inherent significance exists.

On the other side, strong vitalists claim that life introduces a whole new type of causation. Life anchors causal chains, so that one may meaningfully speak of a living being as the first in a chain of efficient causes. They may be moved by mental states or some other immaterial cause, but within the physical world, they are the ultimate cause. Thomas

---

[2] Several philosophical schools deny that physical stuff is substantial (e.g., Idealism). In such systems, "physical" and "natural" would take on different meanings. Here, I will not use the words in those contexts.

Aquinas and Rene Descartes advance this position for humans, Georg Stahl for organisms in general. Henri Bergson and Hans Driesch are among the last to promote strong vitalism among all organisms. Vitalism as this type of agency is out of favor in both biology and philosophy. Ideas of agency as human-specific causation, however, remain pervasive.

In recent decades, philosophers have begun to explore the space in between strict reductionism and strong vitalism. A bottom-up approach attempts to describe living things as a special kind of link in the causal chain. They do not provide an anchor, but they do connect causes in a life-specific way. Often appealing to natural selection or metabolic regulation, they view living beings as emergent causal entities. Proponents include Ruth Millikan and Karen Neander. A top-down approach extends aspects of human-specific causation to other living things. This attenuated agency provides potential precursors for human agency, looking at mental states or primitive mental states in non-humans. Proponents (e.g., David Chalmers and John Haught) attribute traits such as intention and normativity to non-humans.

The middle ground is not new. Emergence and panpsychism both flourished in the nineteenth century. Before the seventeenth century, the line between mechanical and human causation was drawn differently, when it was drawn at all. We now associate the term "soul" with mental states, subjectivity, human intention and agency, and religious ideas of immaterial immortality. For two thousand years, however, the term spanned a much broader range of biological causation and identity. At times, it applied to humans alone, though rarely with all the post-Cartesian modifiers. It was regularly extended to all animals and, occasionally, to all living things.

Activity and Agency

The standard picture of *agency* connects mental states to physical events causally (Hornsby 2004). Commonly, this refers to a human-specific activity. It requires *prospective imagination*, a form of internal representation including images of the future. Agency also requires preferences. The human wants something and, thus, has *appetite, desire*, or—when directed by reason—*will*. Forming a picture of alternate futures, the human prefers one over another and intentionally moves in favor of it.

In the past few decades, philosophers have asked whether this picture is too narrow. Some of these traits may be present independently (e.g., imagination without will), while others may be present only partially (e.g., retrospective imagination or memory but not prospective imagination). Many human activities occur without intention, from digestion, to reflexes, to unconscious behaviors. Many animal activities appear to involve intentions (e.g., tool use by crows). Tyler Burge and Derek Jones suggest "primitive agency" as "characterized by whole-system guidance of coordinated behavior" (Jones 2017, p. xiv; Burge 2009). After defending a form of top-down causation, wherein the organismic unity impacts behavior, both retreat from attributing primitive agency to plants. Invoking sensation or locomotion, they limit primitive agency to animals and mobile microbes. Xabier Barandiaran et al. (2009) suggest "minimal agency" as "an autonomous organization that adaptively regulates its coupling with its environment and contributes to sustaining itself as a consequence." They are more open to extending this type of agency to all living things. Both theories seek to embed a theory of agency within a modern scientific framework.[3]

Martine Nida-Rümelin (2007) divides activities differently, distinguishing intentional actions from both "doings" and "happenings." Agents do the former, though unintentionally (e.g., fidgeting). The latter are done to agents by mechanical processes (e.g., digestion). These distinctions depend upon ontological commitments about whether a subjective entity initiates the causal chain. They also suggest a hard line between active and passive entities. One acts; one is acted upon.

In speaking of causal chains in vegetables, I face a dilemma. If I refer to vegetable processes as happenings or reactions, I run the risk of importing an anachronistic mechanism. Plato saw the vegetable soul as agential, nearly in the modern sense. It has this connotation for many Neoplatonists and antique Christians. On the other hand, if I refer to vegetable processes as doings, activities, or primitive agency, I run the risk of appearing vitalist to modern readers. This is not my wish. Even when I speak of reviving the vegetable soul, I do not see it as immaterial, unnatural, or in any way contrary to modern science. In any case, all these terms seem to preclude the existence of a middle ground.

After much internal struggle, I have decided to begin by using "agency" for causal efficacy in all living things. This extends the term

---

[3] Specifically, they appeal to the "autopoeisis" of Varela and Maturana (1972) and Varela et al. (1974).

beyond the primitive agency described above, so we might call it "proto-agency." Readers may interpret this proto-agency as an enhanced link in a chain of efficient causes or as an attenuated version of standard agency. I intend it as a place holder so that we can explore the range of historical positions between mechanical passivity and standard agency.

Individuality

Most of us take for granted the boundaries of a living thing. We think that we can easily distinguish the outside and inside of any organism. The question is problematic, however, in countless cases. Bacteria frequently digest nutrients outside their cell walls, providing an indoor–outdoor metabolism. They live in diverse colonies, cooperating for both nutrition and reproduction, blurring the line between unicellular and multicellular life. Non-human animals exhibit a range of behaviors that defy boundaries: from tool use to prostheses to agriculture. Humans frequently use the metabolisms of other organisms to promote survival and reproduction.

Many theories of life, particularly souls, provide ways of counting organisms and differentiating them from their environment. They provide tools for naming what is essential to an individual or a species. Whenever we ask whether a trait is internal or external, intrinsic or extrinsic, we have introduced (often implicitly) a position on the individuality of an entity. When thinking about composition, we might ask whether the material aspects of an organism are essential to it, or only accidental. The constant cycling of material components within a body suggested to many philosophers that materiality was essential, but that particular material components were only accidental. When thinking about activity, we might ask whether an organism motivates itself (as the start of a causal chain) or is driven by external factors. When thinking about purpose, we might ask about the interaction of organisms and normativity. Ideas about proper function, purpose, and intent suggest organism-specific, hence individual norms.

Individuality can also be used in a technical sense, as referring to the divisibility or indivisibility of an entity. Many philosophers have asked whether the soul represents a fundamental unit of biology as atoms represent a basic unit of physics.

Purpose

The history of life-concepts cannot be separated from the metaphysics of final causes. How do we make sense of Aristotle's idea that final causes—"that for the sake of which"—define and motivate organisms? For 2000 years, final causes were central to constructions of the soul and thus to explanations of life. The meaning of final cause changed radically over that time.

Several attempts have been made to remove final causes from the natural world, often by making them a form of divine causation, essentially an alternative efficient cause brought about by God's agency. Skepticism and empiricism in the Enlightenment rendered such causes incompatible with natural science. Final causes under this head were ejected from the natural world with the adoption of the mechanical philosophy. It remains unclear where they went. Many were happy to leave them in God's hands, real but unavailable to science. Others tried to stretch science to engage them (i.e., natural theology and intelligent design). Very few tried to eliminate them altogether until the nineteenth century.

Though successful immediately in physics and shortly in chemistry, the mechanical philosophy did not catch on among biologists until Darwin and Mendel provided reductionist explanations of biological function. Nonetheless, biologists continue to use teleological language. Debate persists about how the language of ends maps onto underlying accounts which are truly mechanical. Mary Midgley, Thomas Nagel, and Alvin Plantinga are only the most recent crop of philosophers to ask whether biology is as mechanical in practice as biologists believe. In each of the parts, I discuss how final causes were interpreted in the relevant time period.

PLATONIC AND ARISTOTELIAN APPROACHES

I refer to Platonic and Aristotelian approaches to souls throughout the book. Both attempt to bridge the gap between physical and human-specific causation. Broadly speaking, Platonic approaches favor eternal and incorporeal souls that motivate biological activity. They emphasize the transcendence of biological causes, though they need not exclude them from the physical realm. Souls participate in the greater life of the One, God, or the cosmos. Final causes are more likely to be clearly prospective

and intentional. Aristotelian approaches identify mortal and corporeal processes with souls. They emphasize the immanence of biological
causes, though they need not reduce them to the physical realm. Souls
differentiate both species and biological individuals. Final causes are
more likely to be consonant with physical causes.

Vegetable souls and the theories of life that replaced them address a fundamental issue in ontology and epistemology. How do living things fit
into a larger picture of the world? And how do we account for them in
our explanations? Modern readers will be accustomed to a clear, or at
least intuitive, distinction between thinking things and physical things.
Vegetable life has always existed somewhere in between.

Part I covers the birth of the vegetable soul, indeed the birth of
Western life-concepts in general: vegetable, animal, and rational. Initially,
they operated as a continuum of activities connecting material particulars
to ideal universals. Chapter 2 looks at *psyche* in Homer and the pre-So-
cratics, setting the groundwork for vegetable souls. Chapter 3 reviews
souls in the works of Plato, with an emphasis on the *appetitive soul*, present in all living things. Chapters 4 and 5 cover Aristotle's life-concepts,
focusing on the *nutritive soul* and the activity that defines it. Chapter 6
covers Greek and Roman appropriation of vegetable souls, along with
inputs from Epicureans, Stoics, and Neoplatonists.

Part II covers the development of life-concepts during the Middle
Ages. Dominated by Neoplatonic interpretations of Aristotle, souls
became less material and less physical, but more substantial. I track two
trends. The first led to what I call the Platonic Synthesis. Chapter 7
presents *nephesh*, the Hebrew equivalent of *psyche*, and Philo's attempts
to reconcile Hebrew scriptures with Aristotle and Plato. He introduced the idea of a dual creation—one physical, the other non-physical—that would become central to medieval ideas about life. Chapter 8
explores *psyche* in Christian scriptures, which drew on both Greek and
Hebrew sources. Tertullian and Origen integrated vegetable, animal,
and rational life-concepts with spiritual life-concepts. Chapter 9 deals
with Augustine's recapitulation of Philo's dual creation and introduces
a subjective life-concept. This sets the stage for medieval cosmology
and biology. The human soul, as spiritual and subjective, drew away
from vegetable and animal souls. A second trend, the Aristotelian
Synthesis built on Islamic *falsafah* and the new availability of Greek
texts in the twelfth century (Chapter 10). Thomas Aquinas formalized

this perspective into the system embraced in the High Middle Ages and Renaissance (Chapter 11).

Part III covers the death of vegetable souls in the Enlightenment and the birth of modern science. Descartes and Gassendi embraced the mechanical aspects of Aristotelian causation as well as material and efficient causes. Both ejected formal and final causes from the physical world (Chapter 12). Chapter 13 looks at attempts to understand the relationship between the newly non-physical souls and the physical world, following Bacon, Leibniz, and Kant. Chapter 14 follows subsequent attempts to reject the mechanical philosophy and return to a more Platonic view of life. Early theories of evolution begin to emerge in the seventeenth and eighteenth centuries (Chapter 15). Finally, Darwin introduced evolution by natural selection, connecting organization and purpose, as well as many features of animal and human life, to vegetable activities (Chapter 16).

Part IV traces some of the surprising developments of modern biology, emphasizing challenges to traditional ideas. Chapter 17 looks at the growing tree of life, including protists, bacteria, and archaebacteria. The modern kingdom Plantae represents just one evolutionary trajectory among a variety of "vegetable" life forms. Chapters 18 and 19 explore vegetable and animal life-concepts as they relate to individuality. Chapter 20 asks where this all leaves us when thinking about composition, agency, individuality, and purpose among living things. Should vegetable souls be resurrected?

REFERENCES

Barandiaran, Xabier E., Ezequiel Di Paolo, and Marieke Rohde. "Defining Agency: Individuality, Normativity, Asymmetry, and Spatio-temporality in Action." *Adaptive Behavior* 17, no. 5 (2009): 367–386.

Burge, Tyler. "Primitive Agency and Natural Norms." *Philosophy and Phenomenological Research* 79, no. 2 (2009): 251–278.

Goetz, Stewart, and Charles Taliaferro. *A Brief History of the Soul*. Malden, MA: Wiley-Blackwell, 2011.

Hays, Lindsey, L. Archenbach, J. Bailey, R. Barnes, J. Barros, C. Bertka, P. Boston et al. *NASA Astrobiology Strategy*. Washington, DC: National Aeronautics and Space Administration, 2015.

Hornsby, Jennifer. "Agency and Actions." In *Agency and Action*, edited by John Hyman and Helen Steward, 1–24. Cambridge, UK: Cambridge University Press, 2004.

Jones, Derek M. *The Biological Foundations of Action*. History and Philosophy of Biology. New York: Routledge, 2017.

Marder, Michael. *The Philosopher's Plant: An Intellectual Herbarium*. New York: Columbia University Press, 2014.

Marder, Michael, and Luce Irigaray. *Through Vegetal Being: Two Philosophical Perspectives*. New York: Columbia University Press, 2016.

Martin, Raymond, and John Barresi. *The Rise and Fall of Soul and Self: An Intellectual History of Personal Identity*. New York: Columbia University Press, 2006.

McKay, David S., Everett K. Gibson, Kathie L. Thomas-Keprta, Hojatollah Vali, Christopher S. Romanek, Simon J. Clemett, Xavier D.F. Chillier, Claude R. Maechling, and Richard N. Zare. "Search for Past Life on Mars: Possible Relic Biogenic Activity in Martian Meteorite ALH84001." *Science* 273, no. 5277 (1996): 924–930.

Mix, Lucas J. "Nested Explanation in Aristotle and Mayr." *Synthese* 193, no. 6 (2016): 1817–1832.

Nealon, Jeffrey T. *Plant Theory: Biopower and Vegetable Life*. Redwood City, CA: Stanford University Press, 2015.

Nida-Rümelin, Martine. "Doings and Subject Causation." *Erkenntnis* 67, no. 2 (2007): 255–272.

Varela, Francisco, and Humberto Maturana. "Mechanism and Biological Explanation." *Philosophy of Science* 39 (1972): 378–382.

Varela, Francisco G., Humberto R. Maturana, and Ricardo Uribe. "Autopoiesis: The Organization of Living Systems, Its Characterization and a Model." *Biosystems* 5 (1974): 187–196.

PART I

Birth

CHAPTER 2

Greek Life: *Psyche* and Early Life-Concepts

Living things require special attention. They move more often and less predictably than non-living things. The earliest life-concepts focused on aspects of motion and motivation to account for this unpredictability—or, perhaps, to create a higher-order predictability. Greek theories of life clustered around the *psyche* (ψυχή). Usually translated as "soul," it can also be thought of as a principle of life. The oldest usage comes from Homer, speaking of human life at its limits. Natural philosophers from Thales onward used the soul to make sense of motion in living things.

The five chapters of Part I trace the origin of three life-concepts, the categories most often used to sort different kinds of life and motion. *Vegetable life* covered nutrition and reproduction. *Animal life* covered sensation and locomotion. *Rational life* covered abstract thought. Plato, Aristotle, and their followers attributed souls (or soul-aspects) to each of the three life-concepts, producing vegetable souls, animal souls, and rational souls as well as a comprehensive system for understanding life, a *psychology*. Speculation about life-concepts crystalized into the familiar vegetable-animal-rational division around the fifth century BCE. Greek philosophers cared greatly about persistence and change; how can we identify stable points in a shifting world? What causes one thing to become another? They invoked souls, either as agents of change or as the process itself. They used them to explain why life behaves more dynamically and unpredictably than non-life. This chapter looks at the origins of the soul in Greek thought.

© The Author(s) 2018
L. J. Mix, *Life Concepts from Aristotle to Darwin*,
https://doi.org/10.1007/978-3-319-96047-0_2

One word of warning before we begin: modern concepts of 'body' (as the physical extent of a living thing), 'organism' (as a unit of life), and 'mind' (as unit of reason or awareness) depend on the speculations and theories of these early Greek thinkers. Those speculations did not, however, assume such categories. It would be anachronistic to impose them onto early Greek souls. The authors under discussion were still working them out.

SHADES OF HOMER

Homer's epic poems, *Iliad* and *Odyssey*, provide the oldest known instances of *psyche*. Difficult to date, estimates range from 1100 to 850 BCE. The *Iliad* recounts the events of the Trojan War, while the *Odyssey* covers the long journey of general Odysseus as he returns home. Homer mentioned the soul repeatedly, but only in the context of human death.[1] Tydeus kills the sons of Merops—he robs them of spirit and soul—and Achilles risks his soul in battle (*Iliad* 11.334, 9.322). At death, the soul flees and no longer holds the limbs together.

Homer spoke of a human (or another animal) as a collection of limbs and faculties, not an organic whole. The idea of organism—organs working to a common purpose—must wait for Aristotle. Neither is there evidence of a parallel to the modern English word 'body' (Snell 1953, pp. 1–15). Homer used the word *soma* (σῶμα) to refer to objects and corpses, but never a living being. His word *demas* (δέμας) comes closer, though it seems limited to human appearance or presence. The *psyche* holds human limbs together; when it flies away at death, the pieces fail and fall apart. Similarly, in other animals, the spirit or *thumos* (θυμός) departs at death.[2] The *psyche* has aspects of breath and blood, common analogs of life in Greek thought. Like breath and blood, it is not localized in any one part of the body, but moves throughout, imparting heat and motion. Utter stillness and cold indicate that the soul has left. Homer wrote of the soul both before and after death, but always in the context of mortality.

---

[1] There is one reference to the *psyche* leaving a pig as it dies (*Odyssey* 14.426: τὸν δ' ἔλιπε ψυχή). Fagles renders this as "it gasped out its life...."

[2] The word *thumos* has a connotation of physical breath, though it already captured some idea of life-breath and livingness by the time of Homer.

Concepts similar to consciousness, emotion, and reason required other words. Homer spoke of the heart and spirit of characters moved by hunger, fear, grief, courage, or zeal. He spoke of the lungs and the intellect when referring to wit, thought, and reason. Notions of physical organs intertwine with mental faculties and mental states. This makes it hard for us to match Homer's words to modern concepts.

David Claus (1981) identifies eight words related to life, mind, and personal identity in this period and works through their gradual evolution from Homer to Plato. He includes *psyche* and *thumos* along with *menos* (μένος), *ker* (κῆρ), *hetor* (ἦτορ), *kardia* (καρδία), *phrenes* (φρένες), and *nous* (νόος). Claus shows how each contributed to later uses of *psyche* in Greek literature. *Psyche* gradually took the place of *menos, ker,* and *hetor*. Recurring themes of vital, emotional, and willful activity shaped Classical Greek expectations for what exactly needs to be explained when talking about human life. Claus gathers the various terms and meanings into an idea of *life-force*. This life-force operates as physically manifest energy, as the seat of courage and fear, and as the subject of intellectual and emotional life.

Ancient Greek philosophers did not distinguish between what we now think of as physiology and psychology. Physical, emotional, and mental vitality were all intertwined. Their language did not distinguish between body and mind or between organs and activities. Mental or emotional states appear throughout Homer's works, but he never attributed them to *psyche*. He reserved that word for a whole-self livingness that can be lost.

Homer also spoke of *psyche* when referring to what remains of humans after death.[3] This *psyche* departs from the mouth or a fatal wound and flies to Hades, the underworld. It is clearly much less than the original person. It exists only as the image (εἶδος) or echo of the person who died. In the context of Hades, we usually translate *psyche* as "shade," an incomplete remnant of a once-living being. Modern English equivalents include shadow, specter, and phantom, all of which suggest that the image remains, but without essence or influence.

---

[3]Several scholars have proposed distinctions between the soul that departs at death and the shade that abides in Hades. Snell (1953, pp. 8–9) argues that they are the same, while Lorenz (2009) sees insufficient evidence to judge either way. See also Bremmer (1983) and Crivellato and Ribatti (2007).

The shades of both heroes and villains travel to the same place after death and all of them lack power. They cannot change things either in their own world or in the world of the living. In the eleventh chapter of the *Odyssey*, Odysseus travels to the Western edge of the world and descends into Hades. The sorceress Circe informs him that the shade of Tiresias, alone among the dead, retains his intellect and holds the information our hero seeks. Most shades are nothing but fleeting shadows, unable to speak or even recognize visitors. Some few hold their identity strongly enough to be revived, if only temporarily, by some heat or power within the blood. Living visitors, by offering their blood, can communicate with them. Thus recharged, several shades speak to Odysseus. His mother's shade spells out the rules of Homer's afterlife. When mortals die, the sinews no longer hold flesh and bones together, the spirit leaves the bones, and the soul departs.

Homer's goal was narrative. He told stories of human struggle: risking and losing lives, venturing to the ends of the earth, and trying to get home. The soul gave him a way to talk about humans at the boundary of life and death. Over the next few centuries, the word *psyche* became popular, not only with storytellers, but with religious leaders, doctors, and natural philosophers. They began to use it not only for humans, but for all living things. Its meaning expanded to take on the aspects of unity, power, and action previously associated with specific parts of the body. The soul absorbed the work of heart, lungs, and intellect in dealing with emotions and thought.

Life Before Plato

By the sixth century, *psyche* began to appear in descriptions of plants and animals. Thales of Miletus (c. 620–c. 546 BCE) identified it with motive force, the ability to change one's surroundings. Perhaps the first natural philosopher, he attributed souls to humans, other animals, and magnets.[4] All three "act" on their environment. When you set a rock somewhere, it just sits there. It neither moves nor changes nor interferes with the things around it. Living things are less predictable. Proto-agency—as life-specific causal power—was tied to *psyche* at least from the time of Thales.

---

[4]Cited by Aristotle in *De Anima* 405a19–21.

Humans have *psyche*, in this sense, but we are not alone. Other animals move themselves. Many Classical Greeks included magnets and planets in their description because they, too, move in unexpected ways.[5] Plants are more problematic; they change themselves, but not their location or their surroundings (at least in classical thought). We cannot say how many of these early authors thought plants had proto-agency. Translators render the Greek word *zoa* (ζῶα) as "animals" in English, but "living things" might be more appropriate. Empedocles, Anaxagoras, and Democritus all unambiguously included plants within the *zoa*, though many others did not.

Returning from the Dead

Pythagoras (570–490 BCE) connected all living things through reincarnation. Human life is recycled after death, enlivening animals and plants.[6] He thought we should avoid eating meat and respect certain plants because of this kinship. Anticipating Plato, modern scholars call this transition *metempsychosis* or the *transmigration of souls*. Pythagoras may not have thought *psyche* was involved, however. Nor can we say for certain whether he included plants. We do know that he saw the liveliness of many living things as somehow derived from human life and must look to later authors for more detailed interpretation.

Philolaus of Croton (470–385 BCE), a prominent follower of Pythagoras, clearly distinguished between *nous*, *psyche*, and more fundamental aspects of life, including growth and reproduction (Huffman 2012, 2014). Humans, he said, have all three. Animals have *psyche* (perhaps from a human), but not *nous*. Sensation and emotional disposition pass from life to life, but not knowledge. Plants possess growth and reproduction, but lack both *psyche* and *nous*, and thus are not part of the cycles of rebirth.

Another follower, Empedocles (495–435 BCE) gave us a few more details. He did not attribute persistence to the *psyche*, but to the divinity (θεός) or divine spirit (δαίμων) of an individual. A single divine spirit

---

[5] Planets travel through the sky on a different path than the stars, making them celestial agents. The word "planet" comes from the Greek word for "wanderer" because these objects wander from the perfect circular path of the regular stars.

[6] Burkert (1972), Heinrichs (1979), and Bremmer (1983) argue for transmigration of souls to plants in Pythagorean writings. Huffman (2014) argues against.

can manifest as male or female, animal or plant, as it passes through the ages. For Empedocles, the journey related to punishment and reward. Wicked spirits are sentenced to wander through various incarnations over thousands of years. This suggests that the spirit can retain awareness and knowledge as well as life. He had a higher view than Philolaus of both plants and animals. Still, it is unclear how thoughts and memories could persist through a plant incarnation.

Common Life

Later thinkers expanded the *psyche* beyond Homeric limits. They began to apply it consistently to both plants and animals, either as a derivative form of human life or as a feature of living things throughout the cosmos. It is not always clear which will be the better interpretation. Plato founded his cosmology on reincarnation while Aristotle saw souls as a regular feature of the natural world. Both options were likely available to their predecessors. Either way, *psyche* became associated with a wide variety of activities common to living things. By the end of the fifth century, the word ensouled (ἐμψυχός) was synonymous with 'alive.'

A survey of Presocratic philosophers reveals five life-specific activities associated with the *psyche*: nutrition, reproduction, sensation, locomotion, and reason. Intention also appeared, sometimes associated with locomotion, sometimes with reason. These activities revealed that something was alive.

At the most fundamental level, early philosophers associated life with movement (κινήσεων). We usually think of movement as locomotion—movement from one place to another—but the Greeks had a broader view. They included growth, decay, and alteration (e.g., changing color or changing your mind).[7] Movement was the ability to bring about change: proto-agency.

The Presocratics also associated life with being moved. They saw physical and emotional sensitivity as an internal change in response to external influences. We still speak of things that "move our hearts" or "turn our stomachs." The five senses provide the clearest example, but we might also speak of being moved by fear or wrath. Physical desires such as hunger, thirst, and lust all fall into this category, as do more

---

[7] Aristotle *De Anima* 406a12–14.

abstract desires for justice and freedom. The Presocratics located these desires in the *psyche* or spoke of the *psyche* being moved by them.[8]

Within humans, *psyche* became the seat of consciousness, intention, and reason. The words *phrenes* and *nous* remained popular, but their meaning narrowed to faculties for thought and intellect. The *nous*, in particular, became more abstract and distant from the material world. The *psyche*, meanwhile, started to mean a unified identity, the person having the thoughts and using the intellect (Claus 1981, pp. 49–56).

A three-fold hierarchy slowly emerged. The broadest category, vegetable life, referred to the processes of (or faculties for) nutrition, growth, and reproduction. Animal life included those living things that also demonstrated sensation, appetite, and locomotion. Rational life in humans, was animal life with a mind or reason, allowing it to participate in a more abstract realm of ideas. *Psyche,* or something like it, referred to all three.

Multiple Souls

Common activities created an uncommon problem. If a soul gives nutrition to both humans and plants, and a soul gives reason only to humans, do humans have two souls? Several Presocratics proposed exactly this solution. Some Pythagoreans divided the soul spatially, with mind (*nous*) and reason (*phrenes*) in the head and courage (*thumos*) in the heart. Each was distinct from a life-soul (*psyche*). Philolaus suggested a four-way partition, with mind in the head, soul and sensation in the heart, a faculty for development at the navel, and one for reproduction in the genitals. Other thinkers divided them differently. Democritus, for example, spoke of a distributed animating principle in all living things and a concentrated mass forming the mind in humans.

The idea of multiple souls seems a simple solution until you return to the question of persistence. Many of these authors wanted to know what endured throughout life. We call child, adult, and elder the same person. What do they all have in common? The soul was seen as a unifying factor. If there are many, which one is most essential? Which holds a living thing together through life (and possibly through death)? The

[8] Pindar, Euripides, and Sophocles present *psyche* as subject to hunger and sexual desire (Claus 1981, pp. 73–75). Herodotus and Thucydides speak of courage, justice, and reputation (Lorenz 2009).

next four chapters follow attempts to reconcile multiple-soul and one-soul approaches. Multiple souls can explain multiple activities, but unified souls explain persistence through the changes of life.

THE STUFF OF LIFE

When Homer wrote about *psyche* and *thumos*, or when Pythagoreans wrote about *psyche* and *daimon*, what exactly are these things? Are they material and physical? And how do they interact with other things in the world? For most natural philosophers in Classical Greece, the sensible world was composed of four elements—earth, air, fire, and water. They give tangible stuff its traits. Earth gives heaviness and air gives lightness. Water brings cold and consistency while fire brings heat and motion. The four elements make up all the things we see, living and non-living, and cause changes in them. Greek explanations of necessary (or elemental) motion appealed to the motion of elements. Earth settles down (what we think of as gravity). Fire stirs up (what we think of as heat). These natural *inclinations* (inherent motion) explain all motion in the non-living world.

Some philosophers, the atomists, held that the elements themselves were composed of even simpler entities: *atoms*. The word comes from the Greek for uncuttable; atoms cannot be divided. For atomists, even elemental motion reflected more fundamental motions. Both elemental and atomic theories suggest that the universe is composed of fundamental particles.

Soul Particles

Some philosophers proposed a rarified particle to explain life. They thought that a very fine, very active particle might explain the motion of living things. Something sufficiently refined might pass through grosser matter and animate living bodies. Perhaps the soul was made up of these particles.

Though dualisms did arise, we should not confuse them with the more familiar dualisms of later centuries; the physical/spiritual divide of Aquinas and the mind/extension divide of Descartes remain more than a millennium in the future (Green 1998, p. 160). Greeks were more inclined to think of a spectrum running from gross tangible matter to rarified spiritual matter. Moral and medical philosophers worried about the body corrupting the soul precisely because the two could mix and interact.

For these reasons, souls were often compared to fire, the most rarified and active of elements. Heraclitus, Parmenides, Democritus, and Leucippus all associated the soul with fire. Either souls are made of fire, which gives them their energy, or souls are composed of soul-specific atoms similar to fire. Alternatively, Thales, Hippo, and Critias associated souls with water, noting the aqueous nature of blood and semen.[9] They spoke of life moving in circulation and shaping another generation. Many considered souls to be both material and physical, just as bodies and rocks are. They were simply made of different stuff.

Cosmic Soul

The foremost alternative to soul particles was the harmony theory. There is considerable debate as to what exactly Pythagoras and others intended by this. It could have been nothing more than a simple ratio of elements. Perhaps a soul is three-parts fire, one-part water, and one-part air. Or perhaps a soul is a pattern, as when eight fire particles align to form a cube. Aristotle objected to such simple notions; he thought that they could not bridge the gap from elemental physics to proto-agency. They were too simple to be compelling. More sophisticated interpretations of harmony were relational. Think of the sound made when a string is plucked. The vibration sets up further vibrations in the air, which interact with one another and produce a sound. Psychic harmony might be something like this. In either case, souls come about through relations among material, physical things.

Pythagoras and Philolaus thought that souls participate in a larger cosmic order. They believed that the world fundamentally and holistically follows rules, drawing it toward light or beauty or the good. Along with many Ancient Greeks, they proposed a *logos* (λόγος), a rationality that orders the world. Variously translated as word, account, order, or reason, the *logos* could be compared to a cosmic soul. It holds the world together and moves winds and waters, just as a soul binds bodies together and moves the breath and blood.

---

[9] Aristotle, *De Anima* 405b2–7; *Metaphysics* 983b20.

Different interpretations arose. Some interpreted *logos* as the proper proportions and good ends toward which all things tend, and which humans alone can know through our intellect. Others saw *logos* as an explanation for why things change at all, instead of being in a state of perpetual stasis. With either interpretation, the harmony theory links individual souls to the cosmic *logos*. They are drawn toward it or moved by it. They participate in the order of the world.

If we think of souls in this way, then they are not based on elements, but something else entirely. Something abstract and universal acts through the elements. Such souls are not material, but they might still be physical. In modern physics, the force of gravity is abstract and universal and would draw all the matter in the universe into a dense packet were it not for a competing force causing universal expansion (dark energy). As we shall see, Plato and many of his followers saw the Divine and the *logos* as incapable of fitting into the world of sensible objects. They began a train of thought that would end in the immaterial, immortal souls most familiar to modern readers. Aristotle, on the other hand, proposed something less transcendent. In ancient Greece, both were considered open possibilities.

The early history of *psyche* reveals a world filled with living things that share some, but not all, of our humanity. Living things behave in ways that defy easy understanding. They grow and change. They live and die. They respond to their environment. Finally, humans seem to have an ability to reason, to think about the world around us, to perceive (rightly or wrongly) a fundamental order behind it all.

Homer associated *psyche* with the activity, heat, and movement of living bodies, but only in the shadow of death. He gave us mortal souls that mark the passing of human life. He also gave us shades, the afterimage of human life, fading away in Hades.

The Presocratics expanded the meaning of *psyche* to include a full range of life activities, from nutrition to reason. No unifying theme had arisen, other than a generic view of life. Key questions had started to take shape, however. How do we construct compelling explanations of human, animal, and plant life? How do living things relate to the elements of the physical world? Does biology require something extra? And finally, what holds living things together through life—what persists?

For the rest of the book, we will be exploring the possibility of things bigger than elements but smaller than the whole world. We will be looking at attempts to explain motion that depend not only on the movement of atoms and elements, nor only on the pressure of universal forces, but local agents (and proto-agents) that exist in their own right and at their own level. It was unclear in Ancient Greece, as it is unclear now, that such things are more than convenient labels for understanding the world. Perhaps plants and animals really are just assemblages of elements. And yet, they seem to be something more. They seem to be distinct from their environment and internally motivated. Something strange and wonderful happens to turn elements into a life, just as something terrible and fascinating happens when life leaves the limbs and only flesh and bone remain.

References

Aristotle. *The Complete Works of Aristotle*. Edited by Jonathan Barnes. Princeton: Princeton University Press, 1984.

Aristotle. *De Anima (On the Soul)*. Translated by Hugh Lawson-Tancred. London, UK: Penguin, 1986.

Bremmer, Jan N. *The Early Greek Concept of the Soul*. Princeton: Princeton University Press, 1983.

Burkert, Walter. *Lore and Science in Ancient Pythagoreanism*. Translated by Edward L. Minar. Cambridge, MA: Harvard University Press, 1972.

Claus, David B. *Toward the Soul: An Inquiry into the Meaning of ψυχή Before Plato*. New Haven: Yale University Press, 1981.

Crivellato, Enrico, and Domenico Ribatti. "Soul, Mind, Brain: Greek Philosophy and the Birth of Neuroscience." *Brain Research Bulletin* 71 (2007): 327–336.

Green, Joel B. "Bodies—That Is, Human Lives: A Reexamination of Human Nature in the Bible." In *Whatever Happened to the Soul? Scientific and Theological Portraits of Human Nature*, edited by Warren S. Brown, Nancey Murphy, and H. Newton Malony, 149–173. Minneapolis: Fortress Press, 1998.

Heinrichs, Albert. "Thou Shalt Not Kill a Tree: Greek, Manichean and Indian Tales." *Bulletin of the American Society of Papyrologists* 16, no. 1–2 (1979): 85–108.

Homer. *Odyssey with an English Translation by A.T. Murray, PH.D. in Two Volumes*. Cambridge, MA: Harvard University Press, 1919.

Homer. *Homeri Opera in Five Volumes*. New York: Oxford University Press, 1920.

Homer. *The Iliad*. Translated by Robert Fagles. New York: Viking, 1990.

Homer. *The Odyssey*. Translated by Robert Fagles. New York: Viking, 1996.

Huffman, Carl. "Philolaus." In *Stanford Encyclopedia of Philosophy*, Summer 2012 ed. Stanford University, 1997–. http://plato.stanford.edu/archives/sum2012/entries/philolaus/.

Huffman, Carl. "Pythagoras." In *Stanford Encyclopedia of Philosophy*, Summer 2014 ed. Stanford University, 1997–. http://plato.stanford.edu/archives/sum2014/entries/pythagoras/.

Lorenz, Hendrik. "Ancient Theories of Soul." In *Stanford Encyclopedia of Philosophy*, Summer 2009 ed. Stanford University, 1997–. http://plato.stanford.edu/archives/sum2009/entries/ancient-soul/.

Snell, Bruno. *The Discovery of the Mind: The Greek Origins of European Thought.* Translated by T.G. Rosenmeyer. Oxford: Blackwell, 1953.

CHAPTER 3

Strangely Moved: Appetitive Souls in Plato

Plato (428–348 BCE) used the soul to explain motivations in humans and other life. In so doing, he provided the first clear delineation of vegetable, animal, and rational life-concepts. He observed an internal struggle between bodily hungers, emotional drives, and self-control. This led him to speak of three souls—or three soul aspects, the language is ambiguous—within each person. The appetitive soul, present in all living things, drives desires of the flesh. The spirited soul, present in animals, allows them to move and be moved, both physically and emotionally. The rational soul, present in humans, desires music and poetry, but also works to harmonize all our desires in the pursuit of wisdom. The healthy human and the healthy society value each in proper proportion.

Plato introduced two concepts that remain popular, though we often see them in conflict. Drawing on Pythagoras, he spoke of eternal immaterial souls that come from and aim for a life beyond the present world. These "star-souls" contribute to later theories of souls as supernatural seats of consciousness. Drawing on the natural philosophers, Plato also spoke of mortal material souls that give us biological and social explanations. Through Aristotle, they contribute to physical explanations of life. By the time of the Enlightenment, these two types of souls—indeed these two types of reasoning—will appear incompatible but, for Plato, they were well integrated.

© The Author(s) 2018
L. J. Mix, *Life Concepts from Aristotle to Darwin*,
https://doi.org/10.1007/978-3-319-96047-0_3

FORMS AND PARTICULARS

Plato's souls came in the context of a larger theory about persistence and change in the universe. By the fifth century BCE, two philosophers stood out for their thoughts about change. Heraclitus of Ephesus saw how rivers, seasons, and even humans change dramatically through time. He argued that nothing persists; everything changes and passes away. Parmenides of Elea thought things must persist if we are to know the truth about the world. Truth is simply that which is dependably and persistently real. The continuity of places, concepts, and people suggests an underlying and unchanging order to the universe. How can something come from nothing or vice versa? For Heraclitus, everything changes; for Parmenides, everything persists. This may have been the most important question for philosophers in Ancient and Classical Greece. Neither total chaos or eternal stasis makes sense; reality must be somewhere in between.

Plato's solution divided existence into two parts: the realm of perfect forms, knowable only through the intellect, and the changeable world, which we experience through our senses. A single trait may be attributed to many observations. "Blue" is just such a trait. I can say that the sky is blue, and blueberries are blue, and the coat is blue, and I mean, roughly, the same thing in all cases. Blue is a *universal*, being predicated of many *particular* sensible things. For Plato, the universals were more real and more certain than the particulars that embody them. Particulars fade and change, while the universal remains constant. I will never in my life see a perfectly white object, and yet I can think of perfect whiteness. I use that ideal to judge the whiteness of a piece of a paper or a piece of cloth. There must also be a perfect blue, a perfect justice, and a perfect truth.

The things we encounter with our senses form the realm of particulars. Plato believed they are composed of elements, which are, in turn, made up of triangles. The details sound strange, but perhaps not as strange as modern scientific models. Modern physics, after all, suggests that matter is made up of waves. Composition is a tricky problem. To the realm of particulars, Plato added a second realm of universals. These universals are *substances*, that is things that stand under other things or things that exist on their own. The substantial universals in this realm he called images or *forms* (εἶδος), using the same word as Homer. Plato, however, used it in a very different way. For Homer, the image was less real than the original; for Plato the form is more real, more basic than the particulars that embody it. While we speak of Plato, things that are eternal, unchanging, perfect, and ideal are always forms.

Plato applied Parmenides' concept of eternal, unchanging existence to the forms. He applied Heraclitus' perpetual flux to the sensible particulars. For this reason, Plato doubted that any true knowledge may be gained by observation; our senses only tell us about the temporary particulars. We must rely on our intellect to contemplate the eternal forms. *Philosophers*, "lovers of wisdom," try to overcome the limitations of the particular realm to appreciate formal reality. The theory of forms and particulars plays a crucial role in understanding both what souls are and how we understand them.

Lasting Souls

It would be easy to say that souls are the formal part of human nature and thus eternal, but Plato made a more nuanced move. His most famous doctrine of souls appears in his dialogue *Phaedo*. Socrates has been sentenced to death for corrupting the youth of Athens. His students fear for him, but Socrates is content. Having lived a life of philosophy, he anticipates a good fate. Death, he says, is neither more nor less than the division of soul (*psyche*) and body (*soma*). Philosophers value the pleasures of the soul over the pleasures of the body and are, in life, frustrated by the body's particularity, which interferes with clear thought. The soul persists, though the body does not, because the body is more akin to particulars and the soul to forms. Here we have soul and body paired, but not as opposites. They both exist somewhere in between particulars and forms, being sensible matter in an orderly pattern. The body is more than particular, being coordinated, capable of desire, and persisting for a short time after death. The soul is less than form, being subject to change and tied to particular matter. Nonetheless, the soul, which resembles and perceives forms, must be less perishable than the body. It lasts longer.

The *Phaedo* turns to a reflection on where and how souls persist. Souls differ in the extent to which they are permeated or polluted by the physical. The soul has physical desires that keep it attached to the realm of particulars. As in Empedocles, the souls of wicked humans pass through

Hades to return as animals.[1] Weighed down, they become phantoms in Hades and eventually return to bodies, drawn by physical appetite.[2] Souls overburdened by desire return as donkeys and similar creatures while those bearing too much emotional attachment return as wolves and other "spirited" animals. Even those with social, but not philosophical, virtue return as social insects: bees, wasps, or ants. Only the lovers of learning and wisdom manage to escape the body, which Plato calls a prison for the soul. The ability of bodies to interact with and corrupt souls suggests both are hybrids of the particular and the universal.

In the dialogue, Socrates' students object that this picture of the soul is not good enough. What if the soul lasts longer than the body, but still dies? Socrates' replies satisfy the students, convincing them that the soul never dies. Many modern philosophers have been less convinced. In any case, his answers raise the same issues we ran into at the end of the last chapter. The soul enlivens the body and provides continuity. The soul is also the patient that perceives order and the agent that seeks order in the world; it is a locus of change. Its value as an explanation comes from its intermediate place. Socrates speaks of balancing universal and particular in human life, but other passages in Plato make it clear that souls bring life, harmony, and integrity to other living things as well.

The *Phaedo* ends with Socrates speculating on the fate of human souls after death. He spells out the process of reincarnation in more detail as souls are not only subjected to their own desires but also punished according to their earthly deeds. The most evil are permanently cast into Tartarus, the dark pit below Hades. Others are purified in fiery waters until forgiven by those they harmed. Alternatively, philosophical souls go on to a more rarified life, still on earth but above the air as we are above the waters. Plato made clear, throughout, that these are compelling but not definitive accounts of the soul's fate. They provide a foundation for one of the most popular pictures of immortal souls and an afterlife.

---

[1] Recall that the *daimon* was reincarnated in Empedocles, while Plato is clear that the *psyche* returns.

[2] It is worth looking at the full text, which includes allusions to former soul concepts. "We must believe, my friend, that this bodily element is heavy, ponderous, earthy and visible. Through it, such a *psyche* has become heavy and is dragged back to the visible region in fear of the unseen and of Hades. It wanders, as we are told, around graves and monuments, where shadowy phantoms, images that such souls produce, have been seen, souls that have not been freed and purified but share in the visible, and are therefore seen" (81c–d).

Ruling Souls

Plato wrote about persistent souls in several other places, using various metaphors. In *Phaedrus,* Plato compared human *psyche* to a chariot, with horses and a driver. The horses provide the force to pull the chariot forward, but do not always pull in the same direction. Only the best souls have an intellect (*nous*) capable of managing both and keeping the chariot on track. Plato gave us a hint of his metaphysics, here. That which moves perpetually is immortal; that which can be moved lives conditionally, until it stops moving and dies. To be a self-mover "is the very essence and principle of a soul" (*Phaedrus* 245e).

In *Laws* (894b–899b), Plato equated self-generating motion, life, and ensoulment. He formally defined *psyche* as "motion capable of moving itself" and "self-generating motion." Those things which only move under external force lack souls. With this distinction, Plato opened a divide between the animate and inanimate. Though Plato passed this by relatively quickly, it suggests that he had a distinct psychology which parallels modern biology.

Unlike modern scientists, Plato viewed the inanimate as the exception to the rule. Plato saw the universe as self-moving, and thus ensouled. "The whole combination of soul and body is called a living thing, or animal" (*Phaedrus* 246c). Life and soul apply in the same way to the cosmos, humans, and other animals. The living cosmos contains organs which are themselves alive. Rocks within the cosmos, like a skeleton within a human body, have no life of their own. Still, they are part of the larger life.

In *Alcibiades*, Plato said that the soul possesses and rules the body (I, 129e–131c). From this, he reasoned that the self should be identified with the soul. True identity is *psychic* identity and true love is love of a *psyche*. It would be anachronistic to characterize this *psyche* using modern concepts of mind and personality. It clearly has more of a life-force connotation, and yet these passages argue on the side of a more immortal, more intellectual soul.

Drawing on Plato's ideas of soul as principle of life, that which persists, driver, and ruler, later philosophers identify Plato with an eternal, unchanging soul. They distinguish between the soul—eternal and essential—and the body—corruptable and peripheral. We continue to speak of this as the *Platonic soul*, despite Plato's use of other metaphors. Plato uses the soul to explain the short-term persistence of the body as well as the long-term persistence of the self.

A Continuum of Souls

Plato began to deal with the competing senses of *psyche* by speaking in terms of three distinct souls, or perhaps three distinct aspects of the soul: one for bodily life and appetite, one for courage and drive, and one for reason.[3] A clear line could not be drawn contrasting form and particulars within living bodies; Plato, therefore, created a continuum, running through five stages from elements, through three ascending souls, to the forms and cosmic order.

In the *Republic*, Plato focused on souls as motivators and asked which of our motivations should rule us. Like the charioteer and horses in *Phaedrus*, there are multiple actors, pulling in different directions, but there is also a single actor who sees clearly and steers the collective. Plato spoke of a perfect city with three classes of people (books III, IV, and X). Guardians—selfless philosophical rulers—seek wisdom and harmony for all. Auxiliaries—courageous enforcers—seek honor and repute through maintaining order and defending the city. Farmers and other skilled workers—the bulk of the citizens—seek health, wealth, and physical success. All three help the community, but only when properly balanced.

Human souls are similarly divided.[4] They must be because we experience internal conflict. Something within us seeks wisdom; something else seeks honor, and a third thing seeks success (435d–436a). These desires are distinct and often conflict with one another, as when we recognize a desire to eat or drink to excess.[5] So our souls must be divided,

---

[3] I have chosen to call them three "aspects" of the soul, following Price (2009), though emphasizing their distinct ability to motivate as in Lorenz (2006). Price provides a thorough analysis of whether they should be treated as ontologically distinct.

[4] Plato begins by saying there are two roads to this knowledge (435c–d). The shorter path, arguing from the city and from internal conflict is spelled out and appears in my description. The longer and better path is never fully articulated. Buchheim (2006) explores the topic and we will return to his conclusions below in the section on achieving harmony within the soul.

[5] 436a–441a. The lower desires have no proportion. We desire simply to eat, drink, etc. We have anger or sadness. It is only in our desire for wisdom that we know proportion. Self-control, that is the ruling function of our rational soul, allows us to eat and drink in moderation and to avoid being overcome by grief.

as the city is divided, by competing interests. The appetitive soul (ψυχή ἐπιθυμητικόν) drives physical desires.[6] These include hunger, thirst, and lust. Plato placed the appetitive soul in the belly, below the diaphragm. The spirited soul (ψυχή θυμοειδής) produces courage and anger as well as desire for recognition.[7] The name reflects its relationship with spirit and abstract emotion. Plato said that the heart watches over it in the chest. Finally, the rational soul (ψυχή λογιστικόν) relates to the underlying order of the universe, or perhaps a human ability to find or make order, through reason.[8] It corresponds to *phrenes* and *nous* and resides in the head.

Cities and humans achieve perfection through the proper integration of their parts. In the city, guardians use the auxiliaries to control the populace. In people, the reason directs the spirit in ruling the appetite. Virtue is nothing more than good governance of one soul over the others. Plato ended this section of the *Republic* with an analogy: health is to the body as justice is to the soul. Both reflect the proper alignment of parts. Together, the three souls create a bridge between changeable particulars and perfect ideals.

All living things are material bodies bound together by appetitive souls. They express a proto-agency in feeding and reproducing themselves—vegetable life. Among such basic lives, some have social agency. They recognize not only internal order but the order between souls, the proper relationship of themselves to other souls in their environment. Further, they can act to change these relationships. Beings with such spirited souls have animal life. Among the animals, a few perceive not

---

[6]The appetitive soul has also been referred to as animating soul (Broadie 2001), the function or image of desire (Buchheim 2006), the concupiscible soul (e.g., Crivellato and Ribatti 2007), and simply desire (e.g., Lorenz 2009). The name can be misleading, as all three souls produce desires. Lorenz (2006, p. 46) suggests that it is the strength and lack of proportion of desires in the appetitive soul that warrant its name. I would propose as an alternative, that the appetitive soul, being present in animals and vegetables, is the simplest soul—or simplest aspect of a soul—associated with desire at the most basic level. The spirited and rational souls produce more rarified desires.

[7]The spirited soul has been referred to as the function or image of indignation (Buchheim 2006) and the spirit (e.g., Lorenz 2009). It has also been called the irascible soul. It is oriented to the *thumos*.

[8]Scholars have called the related soul the thinking soul (e.g., Broadie 2001), the function or image of reason (Buchheim 2006), the deliberating soul (e.g., Crivellato and Ribatti 2007), and, simply, reason (e.g., Lorenz 2009). It is oriented to the *logos*.

only relational order, but cosmic order, the very forms of Plato's eternal realm. Humans fall into this last category.

By making the three aspects comparable, Plato allowed them to affect one another and allowed humans to balance the three desires properly. In neglecting the materiality of appetitive and spirited souls, later Platonists preserved an eternal soul, but robbed it of its relation to, connection with, and power over the particular body.

Achieving Harmony

How should we view the three aspects of the soul, if we are unwilling to cleanly divide them into particulars and forms? How is the soul both single and composite while spannng the gap?[9] The soul cannot be both immortal and divisible; divisible things are subject to change in Plato's thought. And yet the soul admits of contrary desires; one actor cannot pull in two directions at once.

Plato did two kinds of work with his rational soul. On one hand, it reflects only one of three types of desire. The appetitive soul rightly seeks food and drink for the body, but food and drink in excess lead to gluttony and drunkenness. The spirited soul rightly seeks physical prowess and acclaim, but too much attention to these can lead to envy and harshness. Similarly, the rational soul seeks harmony, attending to poetry and the study of philosophy, but if we pursue them to excess, neglecting food and physical training, we also fail. There must be enough to make one cultivated and orderly but not soft.[10] Even philosophy, in the sense of loving learning and reflection, must be balanced by the needs of the body and spirit.

On the other hand, the rational soul acts to harmonize all the desires, each of which is inherently good, but only in proportion. Health is a type of harmony in all living creatures. In the *Timaeus*, the soul is created to mediate changeless, indivisible being (that is forms) and

---

[9] For a recent review and multiple perspectives on this perennial question, see Barney et al. (2012). Common themes include a greater understanding of multiple threads throughout Plato (rather than appealing to radical developments in subsequent writings), recognition of a divided self (rather than a unitary control center), and appreciation of what we might call cognitive or psychological components to the appetitive soul.

[10] *Republic* 410d–e. Orderly is κόσμιον and soft μαλακώτερον. Cp *Timaeus* 87e–88a.

changing, divisible becoming (that is particulars).[11] Philosophy has to do with the dynamic ordering of the soul, moving through appetite, spirit, and reason as necessary stages in the harmonization of living matter to the order of the universe. Once we see the soul as the shepherd of becoming, it is less difficult to reconcile the three soul aspects. In humans, becoming is directed to being through the three ascending stages, directed by our perception of forms. The theory provides a compelling link to Pythagorean concepts of soul as a dynamic harmony related to fundamental order. Soul is the overall working out of order, and the soul aspects are individual stages in the process.

WORLD ORDER

In the *Timaeus*, Plato provided a more direct account of his universe. He described the creation of souls. One great Creator set the world in motion.[12] The Creator made the ordered cosmos as well as the gods of the Greek pantheon. The scheme only makes sense once we understand the profound difference between Creator and gods. The Creator stands outside the cosmos and causes it to be. The gods are animals within the cosmos, created and ensouled. This has led some authors to think Plato attributed personality to planets, plants, and places, but it seems more likely he was attributing motion, agency, and organization. Though Plato's word *zoon* is usually translated as 'living thing' or 'animal,' it must not be conflated with later and more restricted definitions (e.g., conscious beings or metazoans). For Plato and many Presocratics, *zoon* included all biological life—and perhaps more.

Plato devoted the bulk of the *Timaeus* to the creation of the changeable world. Using reason (*logos*), the Creator placed intellect in soul and soul in body to make the universe a living thing. The lesser bodies that make up the body of the universe are necessarily composed of fire (to be visible) and earth (to be touchable). Water and air act as glue, holding

---

[11] *Timaeus* 35ab–b. Buchheim (2006).

[12] The word I am rendering Creator is *demiurge* (δημιουργὸς), the same word for skilled worker applied to the lowest class of citizens in the *Republic*. I have labeled it Creator from tradition and in connection with Platonic and Aristotelian notions of the first agent. It matches Christian notions of God only to the extent that Christians adopted Platonic concepts. Note that the *visible*, closest to what we think of as physical matter, already existed; the *demiurge* shapes the disordered visible into *soma*, *psyche*, and *nous* (*Timaeus* 30a–b).

the elements together. The Creator set the cosmos in motion with a soul so that it might spin in perfect circular motion forever.

Within the cosmic body, were created four orders of living things, each corresponding to one of the four elements (38b–40b).[13] The gods are living beings of fire. They are contingently mortal; they could die but will not. The Creator outsourced the fashioning of other living things to the gods, and thus other living things are imperfect and mortal. They include human bodies as well as the birds of the air, the fish of the water, and the walking things of the earth.

The Creator took the leftovers from making the cosmos and formed them into souls in the same number as the stars. He sowed them into time and space as men, giving them sense perception, love, fear, and spiritedness. If they conquered these emotions, they could return to their first home in the stars; if not, they would be reborn as women. Those women who lived just lives might be reincarnated as men. Those who lived unjust lives would return as other animals and so continue in the cycle of rebirth until their star-soul was no longer in thrall to the emotions. This scheme matches closely with the one found in *Phaedo*.

Following the Creator's example, the gods fashioned the four elements into bodies and mortal souls for the motion and organization of living things within the world of becoming. Imitating the Creator, the gods fashioned mortal animals, held together and moved by mortal souls. These mortal souls live, die, and are replaced by new mortal souls for new generations. Here the account of plants and animals gets somewhat confusing. The eternal souls of humans reappear in animals as they pass through their incarnations. At the same time, vegetable and animal souls enliven all living things, even human bodies, but last only one lifetime. It remains unclear exactly how the eternal souls exist within animals, which have a mortal animal soul and no apparent *nous*. As long as we see them as subsistent and independent, the scheme will be confused. It makes more sense to think in terms of a continuum, with individuals moving toward and away from perfect being.

Mortals are imperfect and, therefore, buffeted by the eternal current and impacts from other created things. They are patients, being moved by the world, both physically and emotionally. Plato attributes the senses to this suffering and blames their fallibility on this movement. Humans,

---

[13] See also: *Phaedrus* 246b, *Euthydemus* 302e, and *Laws* 898c–899d.

with all other animals, have such fallible senses. This means that sensation cannot be a function of the star-soul, which is eternal and rational. It must be a product of the lesser souls—appetitive and spirited. The star-souls possess a distinct faculty for knowing, the intellect, which will only be confused by input from the senses. Instead, true knowledge comes from our memories of a time before we were embodied.

In life, the gods separate our souls to keep the lower from corrupting the higher. The rational soul, the star-soul, they placed in the head, separated from the passions by the narrow passage of the neck, with arms, legs, and torso positioned to hold the head away from the ground. The mortal soul was divided into two parts, placed within the trunk of the body and divided by the diaphragm. The spirited soul exhibits courage, spirit, and ambition. It abides in the chest with the heart so that it can command the whole body. The lungs act to cool the heart when it is overheated and beats too fast, a condition caused by spirit or fear, but mediated by the amount of fire in the heart. The appetitive soul desires food and drink. It abides near the navel and the reason controls it through projecting images onto the liver.[14]

Aware of human frailty, the gods gave us protection in the form of plants (*Timaeus* 77a–c). This is the only place in the Platonic corpus where plant nature is specifically addressed, but it makes an important connection. Plants have appetitive souls and deserve to be called living things. They lack opinion, reason, and understanding but possess sensation, pleasure, pain, and desire. These plant souls are literally rooted in place, lacking locomotion. They do, however, regulate themselves and resist being moved, so they have proto-agency and motion in the most limited sense.

The *Timaeus* continues with a discussion of balance, equating both health and virtue with harmony—proper proportions among the parts. When young, a living thing has fresh elements (with sharp triangles) that overcome the elements that it ingests, turning them to the growth of the organism. When old, the worn out elements lose their struggle with external forces, they come apart and eventually release the bonds of the soul. Diseases arise from imbalances or misplacements of the elements. The body can make the soul diseased and vice versa.

---

[14] Likewise, the spleen exists to keep the liver clear and clean, while the intestines slow the passage of food, so that humans need not eat continuously. For commentary on the role of the liver in reason-appetite communication, see Lorenz (2006, pp. 98–101).

Plato provided the first systematic framework for souls as biological reasoning, with a focus on the motivations of humans. Bodies and souls harmonize the perceivable realm with the proper order of the forms. Both represent organized collections of elements. Souls play the greater and more dignified role in actively ordering (or reordering) bodies through nutrition, sensation, and locomotion.

For humans, the soul also acts as the pilot, steering the individual toward escape into the perfect world of being. The desires, however, may diverge, and Plato presents three aspects of the human soul, each of which can move us. The rational soul seeks wisdom and order. It is distinguished by an ability to know forms. The spirited soul is oriented toward spirit and enables sensation, locomotion, and a desire for victory and acclaim. The appetitive soul gives us our basest desires. The rational soul seeks immortality, while the lesser souls are distinctly material and mortal. All three are composed of elements that interact with the elements of the body.

Plants do not participate in the great transmigration of souls, and yet they, too, are organized. They, too, grow and shrink and respond to their environment. Plato suggests that the gods created them as aids to us, possessing appetitive souls, but lacking either spirited or rational qualities.

Plato managed to harmonize the broad range of soul theories. Though he could not eliminate all conflicts between the various views, he made progress toward a unitary and unifying picture of life. His souls exist as a continuum rising from becoming, through organized bodies, into true being. Aristotle appropriated both ends of Plato's continuum—base life functions and unifying harmony—in his own construction of vegetable souls.

References

Barney, Rachel, Tad Brennan, and Charles Brittain, eds. *Plato and the Divided Self.* New York: Cambridge University Press, 2012.

Broadie, Sarah. "Soul and Body in Plato and Descartes." *Proceedings of the Aristotelian Society* 101 (2001): 295–308.

Buchheim, Thomas. "Plato's *phaulon skemma*: On the Multifariousness of the Human Soul." In *Common to Body and Soul*, edited by R.A.H. King, 103–120. New York: Walter de Gruyter, 2006.

Crivellato, Enrico, and Domenico Ribatti. "Soul, Mind, Brain: Greek Philosophy and the Birth of Neuroscience." *Brain Research Bulletin* 71 (2007): 327–336.

Lorenz, Hendrik. *The Brute Within: Appetitive Desire in Plato and Aristotle*. New York: Oxford University Press, 2006.

Lorenz, Hendrik. "Ancient Theories of Soul." In *Stanford Encyclopedia of Philosophy*, Summer 2009 ed. Stanford University, 1997–. http://plato.stanford.edu/archives/sum2009/entries/ancient-soul/.

Plato. *Platonis Opera*. Edited by John Burnet. New York: Oxford University Press, 1903.

Plato. *Five Dialogues: Euthyphro, Apology, Crito, Meno, Pheado*. Translated by G.M.A. Grube. Indianapolis, IN: Hackett Publishing, 1981.

Plato. *Republic*. Translated by G.M.A. Grube. Revised by C. D. C. Reeve. Indianapolis: Hackett Publishing, 1992.

Plato. *Plato: Complete Works*. Edited by John M. Cooper. Indianapolis: Hackett Publishing, 1997.

Price, A.W. "Are Plato's Soul-Parts Psychological Subjects?" *Ancient Philosophy* 29 (2009): 1–15.

CHAPTER 4

Three Causes in One: Biological Explanation in Aristotle

Aristotle (384–322 BCE) reframed the discussion, moving from biological motivators to biological activities—processes in the world. This picture works surprisingly well with modern physics and biology, though some parts still sound magical to modern ears. Causes and organisms dominate in place of hungers and harmonies, but souls still play a crucial role. Like a flipped hourglass, we find the same pieces as in Plato in roughly the same shape but moving in the opposite direction. This resulted in a profound change of perspective. First, Aristotle shifted the focus from universals to particulars. In place of cosmic logos reaching into the world, he gave us a four-fold, particular explanation. Individual things can be understood and known, not just universal forms, and we understand them using four accounts that explain their properties. Traditionally the accounts are called material, efficient, formal, and final causes. Second, he granted all substances both form and matter. Third, and perhaps most important for our consideration, he was genuinely interested to explain life broadly. Plato had provided vegetable souls as the least familiar aspect of humanity, almost an afterthought. Aristotle started with vegetable souls as the principle of life.

Aristotle was first and foremost a biologist. One-third of his works describe or attempt to explain living things. He produced a hierarchy of three souls—vegetable, animal, and rational—as a way to understand how living things behave. Though he flirted with the idea of an immortal intellect, it does not appear to be a priority. This chapter deals with Aristotle's natural philosophy, life-specific activities, and biological

L. J. Mix, *Life Concepts from Aristotle to Darwin*,
https://doi.org/10.1007/978-3-319-96047-0_4

explanations. The next chapter will look more closely at the three souls and how they operate.

A QUESTION OF SUBSTANCE

Aristotle developed a new vocabulary for talking about change and persistence: essences and accidents. Anything we might speak of has essential properties, which cannot change without the thing ceasing to be "what it is." My brain is essential—without it, I would be neither my particular self, nor a living human. Accidental properties, on the other hand, can be changed. I can change my hair color without losing my humanity. Philosophers continue to argue about whether this division helps us understand the world, but it has intuitive appeal and will be central to understanding Aristotle. The fundamental unit of biology, indeed the fundamental unit of existence, will be substances, that is the sorts of things that persist through some change. We can identify a substance by asking what is essential to a thing.

Eating displays this clearly. Why do we say that Sharon eats the cookie instead of saying that the cookie dissolves into Sharon? You can say that Sharon's *essence* persists, while the cookie's essence goes away.[1] Alternatively, you can say we care about Sharon throughout; we only care about the cookie before. It is no longer useful to keep track of the molecules and call them "the cookie."

Substance has both matter and form and the term *hylomorphism* captures this duality; it comes from Aristotle's terms for matter (ὕλη) and form (μορφή). Aristotle also used the word for form found in Plato and Homer (εἶδος), but he used it in a new way. For Aristotle, form captured the essence of the cookie, but only in the context of specific matter.[2] The matter in the cookie is flour, butter, sugar, egg proteins, and chocolate. The form is the cookie shape, but also the role of the cookie as a snack. There are many ways we could put the ingredients together and very few

---

[1]Aristotle might say that the matter loses the form of cookie and takes on the form of Sharon. In this sense, the matter persists, while the forms change. I have chosen to focus on the substance rather than the matter, not because the matter is less important, but because essences will be key to Aristotle's biology.

[2]For simplicity, I am sticking with Aristotle's concept of hylomorphic substances present in *Categories* (and Shields 2007, pp. 53–64). A different concept, embracing forms as substances, appears in *Metaphysics*.

of them make cookies. It was important for Aristotle that the ingredients can exist in another form, perhaps cake, and the form can exist in a different way, perhaps a recipe, but a cookie only happens when they come together.

More importantly, neither form nor matter exists alone. The matter of flour may be found in a cookie, a cake, a clump, a pile, or any number of other forms, but it always has a form. (Medieval scholars introduced the concept of formless or *prime matter*, but Aristotle never spoke of it.) The form of cookie may be present in the flour, sugar, etc. of a given cookie, in ink on paper for the recipe, or as chemical signals in the brain of the chef, but again, it always has some matter associated with it.[3]

Remember this is language for talking about change. Aristotle did not care about which parts are real or even really the cookie so much as he wanted words to differentiate between what we track and what we stop tracking. We follow the substance (form and matter), which stays the same. We don't follow the accidents, like temperature, size, and taste. So, we can say that baking makes the cookie hotter, bigger, and chewier. Only the rare physical chemist follows the heat from oven to cookie or keeps track of one cubic millimeter of space to see when cookie replaces air in it. No one asks where the chewiness was before it was found in the cookie, or after. We care about changes in the cookie...at least until it meets Sharon. Then we care about her and say that the cookie ceases to exist. The form of cookie-ness leaves the matter. Aristotle provides words for understanding why and when we change the subject.

In biology, we can get confused about the subject. Individual plants, animals, even people can be difficult to distinguish from their environment. Eating crosses boundaries, redrawing the lines. When Aristotle speaks of nutrition, growth, and reproduction in biology, he wants to understand how a substance gets bigger by adding more matter to its form. Now that we know what the substance of a living thing might be (form plus matter), let us turn to explanations. How might we understand the existence of a living thing and how do we explain its properties, both essential and accidental?

---

[3] Based on *On the Soul* III.5 and *Metaphysics* Λ, some have argued the Aristotle saw the *nous* as an immortal immaterial aspect of the soul along the lines of Plato. Martin and Barresi (2006, pp. 21–22) and Johnson (2005, pp. 171–172) provide discussion. As our focus will be on the vegetable soul, which lacks intellect, I will not go into greater detail here.

Explaining the World

Aristotle thought there were four important types of explanation. We usually call them the four "causes," but the word *aition* (αἴτιον) can also be translated "case," as in a court case. What evidence do we have for a specific event? What will it take to convince us that we understand it properly? Before we ask how to explain living things, we must look at explanation in general. For Aristotle, all explanations were accounts of the nature or *physis* (φυσίς) of the thing in question—all of them are physics or science in some sense.

Material Causes

One valuable explanation speaks to composition, what a thing is composed of. Being made from iron explains why cannonballs are heavy. Being made from paper explains why books are flammable. And being made with sugar explains why cookies are sweet. Such basic examples give us a good starting point, but composition also explains things in other ways. Material cause explanations ask about parts, but they need not be the most basic parts. Cookies are sweet because they are made with sugar, but not because they are made with carbon, hydrogen, and oxygen, even though those atoms make up sugar. The sweetness comes from how the molecules interact; it cannot be explained at a lower level. We need a different type of explanation to understand why sugar itself is sweet. Nor do material causes need to be tangible, material, or physical. Rather, the material cause captures "that from which" or "that out of which" a thing is made. Letters cause syllables and premises cause conclusions (*Physics* 195a15–17). All are presented as material causes. The material cause of a living thing usually appeals to the organs and systems which are assembled into an organism.

Formal Causes

Another valuable explanation deals with the pattern or shape of a thing. The pattern of metal on the cookie cutter explains why the ginger-bread came out in the shape of a man. The wells in the ice tray explain why the ice is in cubes. The blueprint explains the flow of rooms in a house. Once again, we need to be careful, though, because Aristotle did more with this concept than just speak about the spatial extent of something.[4] He also cared about the organization, the relationships, and the movement of things. His actual term is best translated "the account of what-it-is-to-be a thing" and comes very close to the concept of essence described above.[5] We will get in trouble if we import Platonic notions of ideal, unchanging forms. Substance only happens when matter takes form. Aristotelian forms come and go. When we talk about the formal cause of a living thing, it will be necessary to ask what persists such that we call it the same thing from birth to death.

Efficient Causes

The third type of explanation deals with the actor or means by which something was brought about. A baker explains why there is a cookie… and so does the art of baking. A father and mother explain why there is a child…as does the act of conception. A donor explains why the museum was built…and so do the board of trustees, the architect, engineers, and builders. Efficient causes need not be simple or solitary. Often, they involve a long chain of actors and arts. In Plato, the three aspects of the soul were the efficient causes of human behavior, while cosmic souls were the efficient cause of movement in the universe.

Of the four causes, only efficient causes approach the modern sense of "cause and effect." Indeed, the term "effect" comes from the same root as "efficient." This familiarity can fool us, though. Two types of efficient causes make greater claims, worth thinking about. The *proximate* efficient cause is that efficient cause closest to the effect in question, the nearest in the chain of causes. Consider the game of pool. If I use a cue stick to strike the cue ball and the cue ball hits the eight ball, only the moving cue ball can be counted as the proximate efficient cause of the eight-ball moving. It was the closest cause. Conversely, the *ultimate* efficient cause is the first event in the series, the one that set things in motion. The ultimate efficient cause provides a causal anchor. In this

---

[4] In *Physics* (194b26–29) Aristotle provided several examples, including definition, kind, and ratio.

[5] Johnson (2005, p. 47). The Latin term *essentia*, from which the English word for essence is derived, was coined by scholars attempting to translate Aristotle's phrases τὸ τί ἐστι (what it is) and τὸ τί ἦν εἶναι (what it is to be). See *Metaphysics* VII.4. See also, Cohen (2014).

example, I am the ultimate cause of the eight-ball moving. Questions of agency are wrapped up in how we think of proximate and ultimate efficient causes, each of which must be exclusive—a single first event, a single last event. Our basic concept, however, is much more generic. Aristotle allowed for multiple efficient causes, acting in series or in parallel to make something happen.

Parents give us a way of thinking of many kinds of efficient cause in living things. Mammals have two parents, each of which is an efficient cause. Neither one was sufficient on their own to explain the child; they must cooperate. Those parents had parents of their own, indicating a string of causes stretching out into history. Some of the toughest questions in biology relate to ultimate causes. Where does all life come from? How does a single life begin?

Final Causes

The notion of *final causes* has generated anxiety in philosophy for 2500 years. Myriad philosophers have explained it in myriad ways. Aristotle referred to it as an account of "that-for-the-sake-of-which." The easiest examples come from human intention. Why was the candle snuffed? Because Matthew wanted the room to be darker. A desire for music explains the iPod. A need for trucks to cross the river explains the bridge. This simple intentional final cause is useful by itself, and Aristotle used it repeatedly. The intentional final cause, however, was not Aristotle's only use.

Aristotle cared about change. He wanted to speak of the end state toward which something was moving when the end state played a role in the action.[6] An oak tree is the final cause of an acorn. We rarely attribute intention to either trees or nuts, but we recognize the importance of future generations in the traits of acorns. We might even say that their function in reproduction is essential. Another modern example involves perspiration. We say that animals sweat "in order to" cool off, even though no intention is required.

One might digress here in philosophy of biology. I will admit that the function of an individual acorn or class of acorns may be turned to other ends. Animals eat them. They fertilize the soil when they decompose.

---

[6]For a modern example of end states being necessary to understand causal relations without intentionality, consider contrastivity as discussed in Schaffer (2016).

Similarly, sweating can expel toxic chemicals and send signals to other animals. Still, it makes sense to speak of biological traits as having a "for the sake of which" explanation. The evolutionary biologist Ernst Mayr famously argued that evolutionary explanations require us to invoke final causes, albeit in a restricted sense (Mayr 1961; Mix 2014, 2015). Traits become adaptive if they lead to reproductive success for organisms. Or, to be precise, they are adapted if and only if they led to reproductive success in the past. What it means to say a trait is adaptive in the present tense remains controversial (Sober 1993, pp. 77–87). In any case, we always view biology through this lens of final causes. How do the organs contribute to the health of the organism? Why has the organism survived in this environment? Nearly everything in biology seems to have some sort of "for the sake of which."

Traditionally, philosophers interpreted Aristotle in a way that equated final cause with intent, but in the last 40 years, they have begun to appreciate how much he had other biological ends in mind. He even spoke of final causes in terms of survival and reproduction. In Part IV, I discuss how final causes might be applied in twenty-first-century biology. For now, let us return to Aristotle.

Aristotle was convinced that the best explanations used all four causes, often acting in parallel or one through another. The sculptor (efficient cause) shaped the clay (material cause) into the form of Venus (formal cause) in order to sell the statue (final cause). Not only biological explanations, but all explanations can take this form, if you have enough information. Biological explanations are special, though, because they integrate the four causes in a unique way. In living things, the formal, efficient, and final causes are the same. The essence of an organism is its purpose, and both are inseparable from how it came about. In other words, a living thing can be defined through understanding its source (similar parents) and end (similar children). Aristotle used souls as a kind of explanation unique to living things, where formal, efficient, and final causes coincide.

Biological Activity

Aristotle wanted to explain what motivates all motion. Just as with traits, we observe essential and accidental motions (*Physics* 8.4). Essential motion is intrinsic; it arises from the substance in question, whether that substance is an animal or an element. We might say it moves on its own, without assistance. Accidental motion is extrinsic; it requires the application of an external force. That external force might be the result of human interference (as a dam holding back water), the result of action by another living thing (as the dirt that is pushed aside when a plant sprouts from the ground), or even the result of an imbalance of natural forces (as when water holds a boat up). Elements, Aristotle claimed, have a natural inclination (ῥοπή) to move and yet cannot move themselves.[7] Fire and air incline upward while water and earth incline downward. Because they do not move themselves, they are easily frustrated by obstacles, as with the water pooled behind a dam and the smoke pooled below the ceiling. They stay where they are until the obstacle is removed.

All living things have both natural inclination and proto-agency. Aristotle cites the Presocratics in identifying this natural ability to move the self with soul (*On the Soul* 1.2–3). After discussing numerous options, he identifies the soul with five basic functions. I refer to them as life activities. They are only observed in living things and they are difficult to reduce to abiological explanation.[8] They are nutrition, reproduction, locomotion, sensation and reason. *Nutrition* involves taking matter that is not self and incorporating it into self. More commonly, we speak of eating, using energy and chemicals for sustenance and growth. *Reproduction* involves making copies of self. *Locomotion* involves movement from place to place. *Sensation* involves perceiving forms in the environment, both the physical senses (touch, taste, smell, sight, and hearing) and more emotional senses (e.g., being moved by courage or fear). Finally, *reason* involves having ideas and putting them together to form models and arguments.

---

[7] See Johnson (2005, pp. 131–158) for a review and discussion of whether this involves teleology and final causes. Aristotle holds elements to be primary bodies of which other bodies are composed, but which cannot, themselves be divided (*On the Heavens* III.3). C.S. Lewis (1964, p. 93) has noted that the description of 'inclining' is *less* anthropomorphic and teleological than the modern metaphor of 'obeying laws.'

[8] In Aristotle, they cannot be reduced to elemental motion. In modern biology, it remains unclear how they might be reduced to physics and chemistry. Nowhere does Aristotle provide exactly this list, though two lists come close. *On the Soul* 413a24 "thinking or perception or local movement and rest, or movement in the sense of nutrition, decay and growth"; 414a31 "the nutritive, the appetitive, the sensory, the locomotive, and the power of thinking"; Reproduction appears distinct from nutrition at 415a26 "The acts in which it manifests itself are reproduction and the use of food."

Aristotle believed that one class of explanations could explain all five activities: instances where the formal, efficient, and final causes of a thing are identical. He sees them as active processes in the living world, processes that stem from and lead to themselves in a self-reinforcing way. The next chapter will go into greater detail about how this works. For now, let me sum up by saying that Aristotle identified five fundamental life-activities. He explained each by appeal to a single cause that was the essence of a living thing and "that which brings it about" and "that for the sake of which" it exists. Any time we encounter such a triple cause, we can call it a soul. Thus, Aristotle established a category unique to biology—substances explained by souls—and an explanatory system unique to biology—soul based causal accounts.

For the sake of clarity, I have presented the Aristotelian worldview as a general physics of four causes, within which there are souls—alignments of formal, efficient, and final causes. That makes living things a subset of all things and fits well with modern sensibilities. In looking at Aristotle, it would be more accurate to speak of natural things (both elements and organisms) that have intrinsic ends. They are moved by internal forces. From this starting point, he proposes "mechanisms," things without intrinsic ends that only act in a life-like way because they are compelled by outside forces. Mechanisms are extensions of souls beyond their bodies. A blunter, if slightly less accurate, summary would say that between Aristotle and the present, we have gone from a living universe, in which machines occurred as a product of psychological agency, to a "mechanical" universe in which life happened to arise. Biology was the default for Aristotle as mechanical physics is the default for us.

Aristotle provided a systematic approach to explanations. The world is made up of substances, matter taking form as observable, particular things worthy of our attention. These substances are best explained according to their essences or the properties and motions natural to them. He proposed four specific types of explanations. "Material causes" speak to composition; "formal causes" speak to shape, but also interactions with the surrounding world; "efficient causes" speak to external and accidental influences; and "final causes" speak to "that for the sake of which" a thing occurs. Those categories feel unfamiliar to modern sensibilities and the terms themselves have been freighted with centuries of commentary. Nonetheless, once we look closely at Aristotle's works, they are not as distant from modern explanations as they appear at first.

Aristotle reserved a special place for living things. They possess souls, neither more nor less than the confluence of formal, efficient, and final causes. These souls are not material; they are processes running in material aggregates, turning them into bodies. They are physical, in that they never contradict the elemental *physis*, even though they provide a complementary, biological *physis*. Souls give Aristotle a way to speak about biological causation in nutrition, reproduction, locomotion, sensation, and reason. They are that motion (by definition or formal cause), drive that motion (as agent or efficient cause), and through that motion they perpetuate themselves (as goal or final cause). Only when we see them in all three lights do they make sense, and yet that perspective relies on following Aristotle's causes carefully. Having considered the basic rules of nature according to Aristotle, we now turn to the specifics of vegetable, animal, and rational souls.

References

Aristotle. *On the Soul, Parva Naturalia, On Breath*. Loeb Classical Library. Cambridge, MA: Harvard University Press, 1936.

Aristotle. *The Complete Works of Aristotle*. Edited by Jonathan Barnes. Princeton: Princeton University Press, 1984.

Cohen, S. Marc. "Aristotle's Metaphysics." In *Stanford Encyclopedia of Philosophy*, Summer 2014 ed. Stanford University, 1997–. http://plato.stanford.edu/archives/sum2014/entries/aristotle-metaphysics/.

Johnson, Monte Ransome. *Aristotle on Teleology*. New York: Oxford University Press, 2005.

Lewis, Clive Staples. *The Discarded Image: An Introduction to Medieval and Renaissance Literature*. Cambridge, UK: Cambridge University Press, 1964.

Martin, Raymond, and John Barresi. *The Rise and Fall of Soul and Self: An Intellectual History of Personal Identity*. New York: Columbia University Press, 2006.

Mayr, Ernst. "Cause and Effect in Biology: Kinds of Causes, Predictability, and Teleology Are Viewed by a Practicing Biologist." *Science* 134 (1961): 1501–1506.

Mix, Lucas J. "Proper Activity, Preference, and the Meaning of Life." *Philosophy and Theory in Biology* 6 (2014). http://dx.doi.org/10.3998/ptb.6959004.0006.001.

Mix, Lucas J. "Defending Definitions of Life." *Astrobiology* 15, no. 1 (2015): 15–19.

Schaffer, Jonathan. "The Metaphysics of Causation." In *Stanford Encyclopedia of Philosophy*, Fall 2016 ed. Stanford University, 1997–. https://plato.stanford. edu/archives/fall2016/entries/causation-metaphysics/.

Shields, Christopher. *Aristotle*. New York: Routledge, 2007.

Sober, Elliott. *Philosophy of Biology*. Dimensions of Philosophy Series. Boulder: Westview, 1993.

CHAPTER 5

Life in Action: Nutritive Souls in Aristotle

Aristotle saw nutrition as the most fundamental life activity, the ability to turn not-self into self. That activity undergirds and supports all the others. In a world of change, how can you do anything unless you last long enough to do it? Rocks, of course, last a long time, but they do so by resisting change altogether. Life does something more interesting.

In trying to construct a systematic science of life, Aristotle divided the world into four categories according to the activities of life: mineral, vegetable, animal, and human. *Minerals* include everything that is not alive. They have none of the activities. *Vegetables* operate through nutrition and reproduction. They consume resources and have offspring. The word vegetable inspires thoughts of plants, but the Latin word *vegere* simply means "to be alive." *Animals* interact with their environment through sensation and locomotion. They have the vegetable activities but also sense and respond to stimuli. This gives them something like experience or consciousness. It allows them to feel, desire, and seek. Finally, *humans* reason.

Aristotle gave an additional shade of meaning by layering the activities. Locomotion and sensation allow animals to find and pursue food, so they enhance nutrition. They are sensitive vegetables. Reason allows humans to think through problems while moving in search of food. They are rational animals. This is not to denigrate the importance of reason for reason's sake. Indeed, Aristotle argues that all living things flourish—achieve their greatest good—in the exercise of their highest activity. Humans flourish in reasoning about reasoning. Nonetheless, it is

© The Author(s) 2018
L. J. Mix, *Life Concepts from Aristotle to Darwin*,
https://doi.org/10.1007/978-3-319-96047-0_5

essential to our nature that we can reason about food. Each of the souls occurs in the context of lower-level activities. We think, feel, and move in the context of our vegetable livingness.

LIFE IN ACTION

We have seen that concepts of life and motion are intimately related. Living things do not move like non-living things. They change direction; they act and react; they work toward goals. Some scholars say that Aristotle attributed these properties to souls. They think of souls as essences that exist independently of the changes they cause. For them, souls are like alarm clocks while the functions of life are the bell, a rare but important activity. This interpretation arose in the Middle Ages and dominated Aristotle scholarship in the twentieth century. More recently, classicists have suggested that souls are processes. In this model, the soul resembles the movement of a grandfather clock. It can only be said to run properly if it ticks away the seconds at a constant rate. It must be wound (i.e., energy must be added to make the pendulum swing) but the constant motion defines it, not the rare ringing of the bell.

Aristotle set forth his concept of souls in *On the Soul*. He insisted that souls are things in-action (ενέργεια) and in-fulfillment (ἐντελέχεια). Few words have been as thoroughly debated as these two. They both have something to do with a distinction between activities that may occur and activities that are occurring. Like Plato, there is a sense that the world is in a state of becoming. Aristotle focused on the becoming and claims knowledge about it, not just about the ultimate end of being. He spoke of potentiality (what may happen) and actuality (that which is happening). Matter is the first potentiality of life. The soul activates the matter, making the soul the first actuality of life. This activated thing is not just matter, but an organized body. It is not, however, an organism. Not yet. The soul is also the second potentiality. The second actuality is a soul in action and in fulfillment, the body taking nutrients, the living being. The end, as stopping point, can never be reached. The organism will never be sufficiently fed for all time, only for now. Plants are mortal. They are fulfilled not by satiety but in continuous nutrition. That is their essence. A thing can be both actual and active. For Aristotle, it was precisely this actual activity that characterizes life.

Vegetables are in the process of nutrition. We think of eating as a time-specific event, something we look forward to, start, and finish.

Nutrition is not quite the same. It is the process wherein we use the things we eat to sustain ourselves. Aristotle spoke of taking in matter and heat. Our example, Sharon, uses internal heat to convert cookie-matter into Sharon-matter. She repurposes. In Aristotle's language, she literally re-forms the elements so that they now have her form instead of the cookie form. In modern biochemistry, we might say that the potential energy contained in sugars gets released and stored as reduced molecules. The energy drives metabolic reactions which break down carbohydrates, proteins, and sugars in our food and builds up new biomolecules which we use in our cells. It seems more immediate when we speak of it happening in a human, but this same process appears in plants, as they consume air, water, and nutrients to grow. Even when plants are not growing, they expend energy and consume resources to maintain themselves. Aristotle saw this as the in-action process of nutrition. In his terms, it involves a substance (the organism) persisting through the continuation of a form in a constantly changing set of matter.

So much for action. What about fulfillment? The second word, *entelechia* in Greek, comes from the same root as final cause (τέλος). Later translators came to think of it as "at rest," or having achieved an end. This would make it incompatible with in-action. Instead, we can think of it as in-fulfillment, realized. If the final cause of an organism is to persist, then it is in-fulfillment when the form persists through constantly changing matter. The individual cells of an organism come and go; atoms cycle through; the organism remains.

When Aristotle says that a soul is the union of formal, efficient, and final causes, he means that one process fills all three rolls. Nutrition in-action and in-fulfillment defines organisms; they sustain themselves. It brings them about (they exist through the practice of eating). And it is their purpose (they live to eat). Living things are simply things in the process of actively sustaining themselves.

THE VEGETABLE SOUL

Aristotle speaks of all living things having a nutritive activity (θρεπτικόν) or, as it comes to be called, a nutritive soul. This soul is the organism's identity as a form persisting by repurposing matter. The somewhat confusing language of *On the Soul* (II.1) speaks of potentiality and actuality in one extended metaphor that spells out the relationship.

Aristotle compared potentiality and actuality to matter and form, recalling that substances are formed matter. He was not speaking of formless matter and immaterial form, but of the components that make up a meaningful thing (a hylomorphic substance). In this sense, "matter is potentiality, form actuality" (412a10). So, we can speak of flesh (muscle and bone) as matter being organized by a soul into a body. "That is why the soul [*psyche*] is an actuality of the first kind of a natural body [*soma*] having life potentially in it" (412a28–29). Aristotle explicitly calls the parts of a plant "organs" because they are organized to a common purpose. The soul as body-plan renders disparate parts into a single *organism*. Many philosophers stop there, with the soul being the first actuality of a living thing; they compare the soul to the blueprint for a house, using the traditional metaphor for formal causes.

And yet, this is not enough. Aristotle continued with a second metaphor. We can also speak of knowledge as potential or actual; it is potential when we have learned something, but actual when we put that knowledge to use. More concretely, we can say that eyes give us the ability to see, but we are only actively seeing when our eyes are open. We can think of dead bodies, which are physically organized, but lack the activity of nutrition. They do not sustain themselves, and so they decay. Life is the second actuality of which the organized body is the potentiality. Nutrition in-action and in-fulfillment makes the soul into the efficient cause as well as the formal cause of a living body.

One might argue that the soul is only the second potentiality, and life the second actuality, but this would take away from the soul as essence of a living thing. It would also give lie to saying that it is in-fulfillment.

> Suppose that the eye were an animal —sight would have been its soul, for sight is the substance of the eye which corresponds to the account, the eye being merely the matter of seeing; when seeing is removed the eye is no longer an eye, except in name —no more than the eye of a statue or of a painted figure. (412b18–22)

It follows that nutrition is the soul of a living thing and that the living thing is, in essence, in the process of nutrition. When nutrition is removed from it, it is no longer a living thing, except in name. Though we can speak of a dead body, it is not a body in the sense that a soul is the form of a body, because it is incapable of that activity for which the soul is a form. It is only equivocally called a body in that sense. And so, we see the soul acting as final cause as well. The end of the organism is to be in-action and in-fulfillment as a living soul.

In short, the soul enlivens the body, turning organized matter into organizing matter. The second actuality is the crucial step that gives us a theory of life. It explains why Aristotle felt a need to supplement the elemental nature (*physis*) with a biological nature (*psyche* as *physis* of a living being). Souls do work as explanations of the natural world. They do not contradict the basic physics, but they do explain a very particular process that would not be clear with elemental physics alone (Mix 2016).

Nutrition anchors the soul concept, but it does not complete it. For Aristotle, plants have only this most basic soul and so their highest end and achievement is the fulfillment of nutrition. Other living things have other, higher aspects to their souls.

> This power of self-nutrition can be separated from the other powers mentioned, but not they from it—in mortal beings at least. The fact is obvious in plants; for it is the only psychic power they possess. (413a31–33)

We will return to those other psychic powers shortly, but first a word about reproduction.

Vegetables in Eternity

The most fundamental trait of organisms is their self-perpetuation. Vegetable function captures this in nutrition, which perpetuates the individual, and reproduction, which perpetuates the form beyond the individual. Unlike the stars and planets, which Aristotle described as eternal, all earthly living things must die. The heavens are made of aether, but the local world is made of earth, air, fire, and water, and subject to constant change. Nutrition is not enough to keep a living thing alive forever.

Aristotle moved away from Plato in speaking of hylomorphic substances. The substance of the organism includes both form and matter. That means that the organism dies when the compound is broken up. Vegetable souls are mortal, as are all things under the heavens. Nutrition manages to maintain the individual through small shifts in matter, but the major shift remains. Death comes for us all, even vegetables. There is hope, though. The vegetable soul is also in the process of reproduction, making new copies of the soul in new matter. Such souls will outlive the individual; the form can persist into the future. Such formal eternity (and formal identity) rates below individual eternity (and numerical identity), but it is still better than nothing. The reproductive function becomes lumped in with nutritive function as the central activity of the vegetable soul. Nutrition brings new matter into an existing soul; reproduction copies the soul and incorporates it as a new organism. Within a lineage, all souls are of one form or kind.

These kinds came to be thought of as "species" and form the basis of our modern taxonomy. Aristotle saw them as eternal. Though living things come and go, neither species nor life at large needs a temporal origin.

Aristotle used the word *phyton*, the Greek for "having grown" when he spoke of living things with nutrition and reproduction, but no other activities. It has the same root as *physis* and suggests that growth is not just a property, but an essential nature. This word *phyton* gets translated as "plant" or "vegetable" in English, but "growing thing" would be more accurate.

Spontaneous Generation

Aristotle's picture of kinds traveling through the ages raises an important question. Where do the kinds come from? For the most part, Aristotle was happy to say that they had been present forever. The universe is eternal and so are kinds. Aristotle did not see the origin-of-life and the origin-of-species as meaningful questions. the world is as it always has been and always will be. Plants come from plants. Animals come from animals. Humans come from humans. As it was in the beginning, it is now, and ever shall be, world without end.

Even Aristotle admitted it cannot be that simple. Seeing worms and flies arise from dirt and muck, he believed that they must be quickened from non-living matter. Some simple animals and plants arise by spontaneous generation. He made a brief digression into the topic in the *Generation of Animals* (III.11), while discussing the generation of shellfish. Curiously, Aristotle compared them to plants, stating that plants grow on land and shellfish in the sea. Both are simple and generated irregularly. Shellfish arise spontaneously, that is by the interaction of elements without intervention by a soul. Some plants also arise by spontaneous generation. "Animals and plants come into being in earth and in liquid because there is water in earth, and air in water, and in all air is vital heat, so that in a sense all things are full of soul" (*Generation of Animals* 762a20). Here Aristotle seems to have a slightly different picture of the soul, a physiological concept in addition to his etiological concept (three unified causes).

There is little consensus on Aristotle's physiological concept, but Gad Freudenthal (1995) provides a detailed discussion. He argues that Aristotle's souls are composed of vital heat. Drawing on Presocratic ideas of fire as the basis of life, Aristotle thought in terms of a fifth element, a fire-like substance, more refined than fire and making up souls. This soul stuff is not the aether, with which Aristotle populated the heavens, but is related to it. The vital heat powers life activities, being present in the *connate pneuma* (life-breath) within the blood. Thus breath, blood, and heat are all tied together in the explanation of life. Freudenthal admits that we only have fragments to work from, not a systematic theory. Still, it reinforces the idea that Aristotle saw souls working through necessary causes (elemental physics), though he may not have worked out the details. The heat of the Sun or the heat of the earth stir up elements in the soil, producing short-lived creatures. Aristotle did not dwell too long on the question and it remains unclear how exactly these simple organisms fit into his scheme of souls. Clearly, they have some operation of nutrition as well as locomotion. And yet, they do not exist for-the-sake-of perpetuating themselves. Instead, they arise from necessity. We even get a tantalizing glimpse of Aristotle speculating on the origin-of-life in time (762b30–763a5).

A related question involves how the essence of an organism passes from one generation to the next without the continuity of a soul. Nutrition represents a continuous activity, but reproduction does not. *Generation of Animals* goes into great detail about specific mechanisms of reproduction, but it does not tie neatly into the picture of souls presented in *On the Soul*. Aristotle used farming as his guiding metaphor. Males plant their seed, which carries the form and energy necessary to produce offspring. Females provide only matter for the developing organism. And yet, the male seed has no soul of its own; a soul must be life in action. Instead, it causes the soul of a new organism to arise. The soul produces seed and the seed activates a new soul. Where does the activity exist in the interval? Aristotle never clearly says, sparking intense debate throughout the Middle Ages.

THE ANIMAL SOUL

Beyond nutrition and reproduction, several other activities meet the criteria for souls: namely sensation, locomotion, and reason. The first two become associated with what we might call a second-order soul. Rather than stacking one soul atop another, the nutritive process is wrapped up with other processes in a single form/cause/end. Nutrition is still a part of what it means to be a soul, but that process of in-form-ation makes a type of organism that can do other things as well. Aristotle refers to a sensitive activity (αἰσθητικὸν). Traditionally, this is translated as the animal or sensitive soul.

It is worth noting that "animal soul" is etymologically redundant. The English words animal and soul come from the Latin and Germanic equivalents of *psyche*: *anima* and *saiwala*.[1] The term gives us some insight into the continuing conflict. Both before and after Aristotle, some philosophers spoke of all living things as ensouled. Other philosophers reserved souls to sensitive things, locomotive things, or thinking things. In Aristotle's works, all living things (*phyton*, including plants) have nutritive souls, but only animals (*zoa*) have sensitive souls.

The animal soul does with forms what the vegetable soul does with matter. Keeping hylomorphism in mind, we can ask how animals sense things. Aristotle believed that the pattern of things literally makes an impression on our senses. In some way sights and sounds inform the matter of our sense organs. This in-form-ation arrives through the animal soul, which processes external forms for the good of the organism. It eats forms, though it does not consume them. Instead, the senses provide a way for animals to recognize substances, usually for the sake of pursuing or avoiding them. A rabbit may perceive the form of a fox, and so flee. A fox may perceive the form of a rabbit, and so chase. Nutrition and persistence are fulfilled by the perception of forms.

The idea is somewhat alien to modern sensibilities about substances, but if you think of everything around you as form plus matter you can imagine taking those forms inside yourself, being aware of them. Sensation is that ability for something inside you to be shaped by something outside. As with nutrition, Aristotle carried out the metaphor at two levels (*On the Soul* 2.6). The first actuality of sensation comes in fetal development, when the body is shaped so that it can receive input from the outside

---

[1] In Latin, *anima* also suggests breathing.

world. Touch, hearing, vision, smell, and taste become possible through an organization of matter in the body. The second actuality of sensation comes when the properties of the external world impact the soul.

Sensation cannot be separated from desire, for whatever the soul senses comes with pleasure or pain, according to the primary function of the soul—nutrition. The soul perpetuates itself. Nourishing food tastes good, and so the animal desires it and seeks it out. Nourishing sensations likewise please the animal and the animal pursues them. Because the highest good of the animal includes sensation, it not only maintains (and passes on) an organized body; it maintains and passes on a body capable of sensation. Self-awareness need not be part of the process. Rather, the animal by its nature avoids what is bad for it and seeks what is good. Pain and pleasure are identical to the fulfillment and frustration of the animal's proper end. Nutrition has been extended from simple eating, to sensing and seeking out things to eat. Reproduction has been extended from simple copying to sensing and seeking opportunities to replicate. It helps here that Aristotle thought primarily in terms of sexual reproduction, which requires finding a mate.

In pursuing their ends, different animals have different types of sensation. Touch is the most basic; Aristotle considered it fundamental to interacting with the environment. An animal reacts to forms present in adjacent substances.

Aristotle recognized a gradation among the animals from the simpler (having only touch) to the more complex. Locomotion, present in some but not all animals, allows the pursuit of desired objects, and avoidance of pain. In this way, locomotion can be compared to the most basic levels of willed motion and intentional agency.

Plato spoke of appetitive souls in all living things. Aristotle thinks this is too much. Broccoli doesn't crave nutrients in the way we crave broccoli. Plants have a good—they need nutrients—but only animals can actively seek their good. Aristotle associated appetite with sensation. Plants passively receive nutrients but they neither act nor react to achieve their good. Animals take in the forms around them, finding them pleasant or painful. They desire the pleasant and have the power to pursue it.[2]

---

[2] Hendrik Lorenz (2009) discusses whether this entails cognition of some sort in animals or even plants. It clearly does not entail reason, self-consciousness, or abstract thought, but it may require associating means and ends, if only to the extent that the living thing connects an activity with pleasure prospectively.

THE RATIONAL SOUL

Aristotle described an even higher order of life in the form of rationality. Some animals have a mind (*nous*) capable of contemplation. To this end, they have a rational activity (νοητικὸν). As with the other two souls, this soul is in-action and in-fulfillment. The rational soul does not simply grant the power to think; it is the continuous process of thinking. As with the initial metaphor, the first actuality of the rational soul is the organization of matter into a body capable of thought. The second actuality requires active thought.

Here Aristotle departed slightly from his scheme of souls in bodies. The forms acquired through sensation always occur with matter, both the matter of the thing observed and some form of matter within our sense organs. In thought we deal with forms that exist only potentially; they are not actualized. Thus, thinking is not about material things and it may not be actualized materially or physically. The rational soul, then, allows organization and action independent of matter, body, and nutrition. It may exist before and after the lifetime of the other soul activities. The rational soul is not necessarily mortal as the vegetable and animal souls are. Having made such an aside, Aristotle goes on to treat the rational soul as a third tier of the life process described so far. It plays itself out in our pursuit of food and mates in a more abstract fashion. More importantly, for Aristotle as for Plato, nutrition and sensation aid us in coming to a life of contemplation. Fulfillment of a plant is in nutrition and reproduction. Fulfillment for an animal is in sensation and locomotion. But fulfillment for a rational animal must be in contemplation, reasoning about reason. Aristotle aimed for a life defined by reason, effected by reason, and aimed at reason. Not only is our highest good a life of contemplation, it is a life contemplating contemplation itself. Thus reason, like nutrition and sensation, is a formal, efficient, and final cause process.

HIGHER AND LOWER ANIMALS

Aristotle could not subdivide the vegetable soul. Nutrition operates in a thing or it does not, leaving us no middle ground. It is alive or it is not. The animal soul, however, is more ambiguous. Aristotle's biological works describe a huge variety of organisms and he intends the three-tiered system to be descriptive, rather than strictly definitive. Animals have a huge range of functions from very nearly plant to very nearly human. Even reason may not provide a clear boundary.

On the Soul contains several chapters devoted to activities found in some animals but not others. Aristotle attempted to describe how they operate. His accounts do not always hold up in the light of modern knowledge, but they do show us how he thought about the relationships between animal kinds. Lines can be drawn between the bloodless and the blooded, between the breathless and the breathing. Both have something to do with the circulation of heat in the body, which allows different kinds of motion. Both appear to be steps toward higher function. For example, breathing animals may have voice, allowing them to make sounds with meaning (420b–421a).

Another highly debated concept in Aristotle deals with the imagination (φαντασία). The word refers to some use of forms, uncoupled from sense perception, including memory and anticipation. Memory, or retrospective imagination, retains sensed forms after they are no longer sensed. Anticipation, or prospective imagination, crafts forms before they are sensed. Lorenz (2006, pp. 124–137) argues that Aristotle saw these as necessary for animals to desire and seek their own good. Aristotle stated that "in all animals other than man there is no thinking or calculation, but only imagination" (433a10–13).[3]

Finally, it is worth noting that Aristotle introduced reason as a human activity but allows that rational souls may be present in another kind of animal (414b19). We may not be alone in our rational animality.

The nutritive soul forms a foundation for all life in the processing of matter to perpetuate kinds. This may be the first systematic account of life with both etiological and physiological aspects. Etiologically, life can be explained as the confluence of formal, efficient, and final causes, also known as souls. The nutritive soul, present in all life, is defined by, brought about by, and aimed towards the incorporation of matter. It literally forms a living body from available elements. The sensitive soul, present in organisms that sense and move, is defined by a similar acceptance of outside forms into the body. It allows animals to actively move and seek resources. The rational soul, present in humans, is defined by its transcendence of the physical processes in which it is firmly embedded.

---

[3] "thinking" here relates to *nous* and "calculation" to *logos*.

Physiologically, Aristotle gave us a partially formed theory of vital heat, the energy by which soul processes operate. Present in small concentrations in nature, this vital heat is concentrated and refined in the life-breath of organisms and mixed with the elements. Whether or not we accept that picture of Aristotle's physiology, we must recognize his profound desire to place souls concretely in the context of matter and physics. His souls are embodied (physical, not material) and active.[4]

Aristotle's physiological account of vegetable souls is not compatible with modern biology. Likewise, the animal and rational souls will introduce vexing problems of interpretation once we ask how to speak of interiority (for sensation) and abstract thought (for reason). The etiological concept of vegetable souls carries no such baggage.

Aristotle provided a theory, perhaps the first, that allows us to individuate living things as "organisms." They have ends toward which they can be organized. His souls became the dominant theory of life in Europe for the next two millennia. Philosophers of all stripes invoke mortal natural souls (or some equivalent) in concert with matter to explain the life of vegetables and animals, even the bodies of humans. And yet, the Platonic notion of immortal, intellectual souls, which Aristotle noted in passing, remains as well. Within a 100 years, students of Aristotle begin bending his etiological theories to make them less physical in order to support human persistence and organismic participation in the *logos* of the cosmos. The next few chapters turn to early attempts to integrate Platonic and Aristotelian life-concepts.

References

Freudenthal, Gad. *Aristotle's Theory of Material Substance: Heat and Pneuma, Form and Soul.* Oxford: Oxford University Press, 1995.

Lorenz, Hendrik. *The Brute Within: Appetitive Desire in Plato and Aristotle.* New York: Oxford University Press, 2006.

Lorenz, Hendrik. "Ancient Theories of Soul." In *Stanford Encyclopedia of Philosophy*, Summer 2009 ed. Stanford University, 1997–. http://plato.stanford.edu/archives/sum2009/entries/ancient-soul/.

Mix, Lucas J. "Nested Explanation in Aristotle and Mayr." *Synthese* 193, no. 6 (2016): 1817–1832.

---

[4] Aristotle's souls are physical, obeying the same rules as tangible things. They are not material, but can only be in-action and in-fulfillment in the context of a material body.

CHAPTER 6

Plants Versus Animals in Hellenistic Thought

Aristotle's causes and souls were slowly reinterpreted over the following centuries. They influenced biological thought for over 2000 years but started to change almost immediately after his death. Aristotle's own life-concepts were not the same as the "Aristotelian" life-concepts of the Middle Ages. The language of souls took on a more technical character as philosophers attempted to create systematic theories of life. Three major changes stand out. First, thinkers questioned the continuity of vegetable and animal souls. The vegetable/animal divide became more significant than the animal/rational divide. Second, numerous schools attempted to unite the best aspects of Plato and Aristotle, often reinterpreting one or the other in radical ways. Epicureans, Stoics, and Neoplatonists formed their own versions of the three-soul theory. Physicians began work on a systematic anthropology useful for anatomy and medicine. All of them continued using familiar terms but recast them in new ways. Four of the most important were, passing from Greek to Latin to English, *physis-natura*-nature, *psyche-anima*-soul, *pneuma-spiritus*-breath, and *nous-mens*-mind. Third, Neoplatonists emphasized the participation of souls in cosmic life and de-emphasized individuality.

The five life activities remained central in philosophy and medicine. Scholars of life focused on the language of souls, with a tension between division and unification. Nutrition and reproduction were closely related, as were locomotion and sensation; reason required something extra. Physicians and Stoic philosophers began to make stronger distinctions

© The Author(s) 2018
L. J. Mix, *Life Concepts from Aristotle to Darwin*,
https://doi.org/10.1007/978-3-319-96047-0_6

between accounts of vegetable and animal activities, with the soul being more closely associated with the latter. After centuries of strong materialism in Epicurean and Stoic circles, the Hellenistic world shifted toward a kind of vitalism under the influence of Neoplatonism. Along with changes in epistemology, this set the stage for medieval theories of life.

THEOPHRASTUS

A brief mention must be made of Theophrastus (371–c. 287 BCE), a student of Aristotle who succeeded him as the head of the Lyceum. While Aristotle focused on animal life, Theophrastus described plant life (Hall 2011, pp. 28–38). Like Aristotle, most of his work was descriptive. Sadly, his treatise, *On the Soul*, has been lost. Other works make it clear that he attributed greater dignity and function to plants than did his teacher. He claimed that they have autonomy, sensation, and their own patterns of growth. While this influenced contemporaries, he dropped from sight before the Middle Ages and had little impact on later thinking.

EPICUREANISM

From the third century BCE to the third century CE, two schools of thought dominated Mediterranean philosophy: Epicureanism and Stoicism. Both built on Aristotle and favored material souls. Epicureanism got its name from Epicurus of Athens (341–270 BCE). Most of his work has been lost, but a summary can be found in his *Letter to Herodotus*, which sets forth a general ontology, including an account of souls (Epicurus 1993, souls are discussed at 35–38).

For Epicurus, the universe was composed of infinite atoms moving in infinite space. Everything of interest, including souls, has a body made of atoms. Biological functions arise through one body acting within another. Fine, air-like particles make up the first; earth and other particles make up the second. One, rarified body, the soul, permeates a grosser, tangible body. The corporeal soul relates to affection, sensation, and movement (von Staden 2000). Epicurus did not go into detail about how this works or how he divided living things into groups.

The next most prominent Epicurean, Lucretius lived in the first century BCE. His great work, *On the Nature of Things*, deals extensively with life and souls (Lucretius 1922, books three and four). He invoked a special kind of soul atom, or soul seed, which combines with wind, air, and fire to make a corporeal soul. This soul is so fine that it has no detectable mass or volume. It disperses at death and, though it leaves the grosser body, it cannot exist on its own. Both Epicurus and Lucretius denied life after death in any form.

Souls make all the life activities possible. Lucretius differentiated the feminine *anima* from the masculine *animus*.[1] The *anima* pervades the whole body, allowing for nutrition, sensation, and movement. The *animus*, located in the chest, moderates the passage of thought atoms and empowers reason. In humans, Lucretius said neither could exist without the other, but he also spoke of animal, and occasionally, vegetable souls. Meanwhile, worms can arise within corpses by spontaneous generation, reusing their soul atoms but not their souls. Heredity also comes from the recycling of soul atoms.

Epicureanism was suppressed by ancient and medieval Christians for several reasons, including divine indifference and the mortality and corporeality of human souls. Pierre Gassendi and Thomas Jefferson would both draw heavily on Epicureanism during the Enlightenment. They would not, however, continue Epicurean life-concepts.

PHYSICIANS

Hippocrates of Cos (c. 460–c. 375 BCE) was among the first prominent physicians. Many works bearing his name helped to shape the Greek medical tradition. Regardless of authorship, they provide some indication of early medical life-concepts (Gundert 2000). They set the tone for later models, stating that body (*soma*) and soul (*psyche*) are both made of the same matter. The body is a dynamic system of fluids, coordinated to a common end. The soul is rarely discussed but operates through the body. It can refer to a generic life principle or, more specifically, to a kind of inner life, associated with the mind (*nous*).

Three further physicians strongly influenced the development of life-concepts in antiquity: Herophilus, Erasistratus, and Galen. Each invoked a corporeal soul and distinguished between life-functions and soul-functions. Heinrich von Staden (2000) discusses all three, exploring the relationship between body and soul. I cover the first two here and return to Galen at the end of the section on Stoicism.

---

[1] *Anima* is often translated as 'spirit' and *animus* as 'mind' but they should not be confused with the Greek *thumos (Latin: spiritus) and nous (Latin: mens)*.

Herophilus (c. 335–c. 280 BCE) lived in Alexandria during a brief period when dissection of human cadavers was allowed. With a strong interest in human nerves, he agreed with Epicurean ideas about a corporeal soul. Unlike them, he thought that it was composed of breath. He distinguished between the basic life-functions (nutrition, etc.) and the soul-functions of sensation and willed locomotion. Explanations of the former require only nature, while explanations of the later require a soul. This distinction caught on through the writings of Chrysippus and later Stoics.

A contemporary, Erasistratus of Ceos developed a reputation for studying human nerves and veins. His position, *pneumatism*, held that all life is based on breath. When breathing, we inhale common air. The heart refines the air, producing life-breath (πνεῦμα ζωτικόν) which circulates in the veins alongside the blood. Basic life functions are mediated by the blood, which carries both nutrients and life-breath. Within the brain, life-breath is further refined to form a soul-breath (πνεῦμα ψυχικόν), which in turn facilitates perception, will, and cognition.

This differentiation between life-function and soul-function set the foundation for later divisions between physicians (working with the *physis* of living things) and psychologists (working with *psyche*). These words remain far in the future, but Hellenistic medicine set the stage for a hard divide between two approaches to life.

STOICISM

The Stoics developed corporeal theories of the soul along the lines of pneumatism. Founded on the ideas of Zeno of Citium, Cleanthes, and Chrysippus in the second century BCE, very little remains of early Stoic writings. Their doctrines must be reconstructed based on later evidence. Von Staden (2000) argues that the Stoics were not dualist, but viewed both the universe and living beings as ensouled material continua. Like Plato, they understood biology as souls working through living matter. Lesser lives participate in the greater life of the cosmic animal. Like Aristotle and the physicians, they saw the breath as a fine corporeal entity that permeates a grosser tangible body.

Some bodies are inherently unified; they are meaningful wholes because of their breath or *pneuma*. That *pneuma* is qualified by its tension (τόνος). More complex bodies with more diverse activities have more tension. At the bottom of the scale are simple, coherent objects, such as rocks. Their *pneuma* forms a *hexis* (ἕξις), often translated as state or disposition. The *hexis* holds a body together but does little else. *Pneuma* with greater tension provides bodies with an internal principle of motion, the kind of motion or activity Aristotle associated with souls. Plant *pneuma* forms a living-breath, literally the nature-breath (πνεῦμα φυσικόν). In addition to holding plants together, it allows nutrition, growth, and reproduction. In animals, the soul-breath (πνεῦμα ψυχικόν) allows motion and sensation. Humans, the rational animal, have an even tenser *pneuma* that forms a ruling faculty (ἡγεμονικόν). The ruling faculty is located in the heart, the hottest part of the body, and manages the whole through circulation. Similarly, the universe contains a ruling faculty, housed in the hottest location, the Sun, and motivating circulation.

The Stoics recreated Aristotle's hierarchy of souls but renamed it as a hierarchy of breath. They perpetuated the human, animal, vegetable, mineral typology, but reinforced the break introduced by the physicians. Souls now refer primarily to the first two categories, while the latter two can be described in terms of their natures alone. This division contributed to dualistic thinking about souls and bodies in later thought. They associated the former with human identity. Later Stoics asserted that all animals grow as vegetables while still inside their mothers. The living-breath is sufficient to explain embryonic growth; ensoulment proper only occurs at birth.

Galen was a physician and philosopher living in Pergamum and Rome in the late second century (129–c. 200 CE). Though human dissection was illegal, he dissected other animals and tried to reconcile Platonic life-concepts with the physicalism of the physicians, Stoics, and Skeptics. With an enormous body of work, it is not clear that a single theory can be attributed (Singer 2016). He presented numerous opinions and, at one point, admitted to speaking of plant "souls" when among Platonists and plant "nature" when among physicians (*On the Natural Faculties* 1.1, Galen 1963). In other works, he specifically associated Plato's rational soul with the brain and nerves, Plato's spirited soul with the heart and arteries, and Plato's appetitive soul with the kidney and veins. With the Stoics, he invoked *pneuma* for all three, but unlike them, he asserted that it is neither the substance nor the dwelling place of souls. Rather, *pneuma* is their instrument. He saw souls working through the material components of an organism.

NEOPLATONISM

The third century witnessed a major shift in life-concepts with the rise of Neoplatonism. As the name suggests, this new philosophy moved away from the physical particular focus of Aristotle and toward the holistic agential focus of Plato. The founder of Neoplatonism, Plotinus (204/5–270 CE) traveled extensively in the Greco-Roman world but lived mostly in Alexandria and Rome. His thoughts on souls were gathered in a comprehensive work, the *Enneads*.

The third and fourth Ennead place the soul in the context of a larger cosmology (esp. 4.7–9). The cosmos emanates from a single, divine unity, the One.[2] Its existence and orderliness come from the unifying dynamism of the One; in other words, the cosmos is biological. Plotinus spoke of the first emanation as the Intellect, which holds and contemplates the Platonic forms. The second emanation (through the Intellect) is the Soul. It plays out the life of the world in matter, primarily through living beings, and more broadly in physical nature. The whole cosmos participates in the One through the Soul and the Intellect.

Aristotle's souls were ontologically stacked, starting with nutrition. Plotinus worked in the opposite direction. His starting point was the ontological primacy of the One, intelligible as Intellect. The cosmic One unfolds in the cosmic Intellect and cosmic Soul. The intellect and soul of smaller living beings also unfold, participating in the grand emanation. Just as human tissues are difficult to understand in modern biology without reference to the function of an organ, so all material things in Neoplatonism are difficult to understand without reference to their function in the cosmic organism. Even rocks participate in the living cosmos (as nails and hair participate in a common organism).

Plotinus used Plato's organic dynamism (becoming moving to being), but framed it as an Aristotelian account (focused on explanatory priority). Thus, a very similar perspective, with souls moving matter into the cosmic harmony, takes on a different language: cosmic unity unfolding into the material world. Both frame souls as a participation in the

---

[2] Gerson (2014) argues that "derive" captures the process better than "emanate," since it is neither an unpacking nor a differentiation in time. Rather, with Aristotle, Plotinus is concerned with the ontological dependence of one thing on another and the human process of understanding or accounting for that dependence. I think this is an important distinction, but retain the more common term.

organic and organizing universe. Both frame vegetable souls as an extended and distant part of this processes, less harmonized than animal and rational souls. And yet, Neoplatonism has a distinct, Aristotelian language.

Plotinus wrote, "In a certain sense no doubt all lives are thoughts – but qualified as thought vegetative (φυτικὴν), thought sensitive (αἰσθητικὴν) and thought psychic (ψυχικὴν)" (*Enneads* 3.3.8). Here, the soul has been identified specifically with the rational function, but in the context of cosmic Soul and Intellect. Meanwhile, the cosmic life is repeatedly compared to the life of a tree (e.g., 3.3.7). Throughout, he dealt with the now-familiar problem of unifying life-concepts to include vegetable, animal, and rational while preserving distinctions between the three.

> And here we have the solution of the problem, 'How an ensouled living form can include the soulless': for this account allows grades of living within the whole, grades to some of which we deny life only because they are not perceptibly self-moved: in the truth, all of these have a hidden life; and the thing whose life is patent to sense is made up of things which do not live to sense, but, none the less, confer upon their resultant total wonderful powers towards living. (4.4.36)

After exploring the life-concepts of previous philosophers, Plotinus attempted to reconcile the rational, animal, and vegetable aspects of the soul. He argued that there are not multiple souls, but a single Soul. Every living being is alive only through participation in the life of the One. Just as we might speak of a soul having multiple activities in a single body, so we can speak of different living things, including plants, as the multiple activities of one cosmic soul (4.9.5).

Plotinus strongly rejected the corporeal theories of his predecessors. For the same soul to permeate and motivate the whole body (be it singular or cosmic), it cannot be spatial in any meaningful way (4.7.8).

Neoplatonism continued the threefold hierarchy of vegetable, animal, rational, identifying soul alternately with the rational component and the overarching movement of life in the universe. Plotinus was not a substance dualist; he advocated for a universe that is both organismic and developmental. The One constantly incorporates unformed and unfavored material into its life, much as we incorporate food into our bodies through nutrition. Bearing this in mind, Neoplatonism seems a less radical shift than it might at first appear. It answered the same questions as Plato, Aristotle, Epicurus, and the Stoics, but framed them a new way. It emphasized the immaterial and nonphysical character of souls as well as their special significance for human life. Vegetable souls remained the primary concept for understanding all living beings, but they moved farther from Aristotelian life-concepts as they moved closer to cosmic life.

Aristotle Reinterpreted

While Greco-Roman thought was moving the soul concept away from the physical and away from plants, it was also rethinking Aristotle's causes. The final cause began to shift at least by the second century CE and slowly migrated from a generic end, or "that for the sake of which," to a more specified intrinsic and intended purpose. Aristotle's explanation had been somewhat obscure, leading to divergent explanations. His clearest examples invoked human intention leading many readers to see that as the only kind of final cause.[3]

Because souls were a concordance of formal, efficient, and final causes, changing final causes also changed souls. Following the rise of Neoplatonism, it became easy to think of souls as laden with the thought and intention of the One, or in the words of Christianity, the mind of God. Philosopher Monte Ransome Johnson (2005, pp. 15–39) summarizes this change in perspective in his book on Aristotelian teleology. Early commentators tried to unify Aristotle's biological and abiological explanatory accounts of "in fulfillment." It came to be interpreted as fore-ordained by cosmic Intellect for the Neoplatonists and by Divine Providence for later Christians. This contributed to a view of the universe in which the One and souls, both immaterial, cause material events. Theologians in Islam, Judaism, and Christianity were all heavily influenced by this Neoplatonic interpretation of Aristotle. Chapter 10 looks more closely at this process and the construction of natural efficient causes and divine final causes as parallel and competing explanations.

Greek and Roman thinkers synthesized and reimagined Platonic and Aristotelian life-concepts resulting in a radically new view of biology:

---

[3] Allan Gotthelf (1976) argues that Aristotle intended his final causes to be empirical in character—observable potentialities and ends. They were neither divine intention (as Aquinas contended) nor a priori tools for understanding nature (as Kant contended).

participation in cosmic life. The developmental and agential aspects of Platonic souls were melded with the explanatory accounts of Aristotle and awkwardly fitted to the material world. The Intellect and intention of a cosmic life flow out to all matter through a hierarchy of souls.

Philosophers clearly distinguished mineral, vegetable, animal, and rational beings as discrete categories with unique explanatory accounts. Each occurs nested within the previous category but possesses its own activities. The significance of this hierarchy was different in different circles, but the hierarchy remained the same. For the Stoics, it was a materialist hierarchy of increasingly rarified breath. For the Neoplatonists, it was a hierarchy of fuller participation in the One. The Stoics approached material, physical monism, while the Neoplatonists approached ideal monism, but both provided a continuum. Souls in this period must, then, be thought of as the movement of material components toward perfection or the movement of ideal Intellect toward matter.

Antiquity provides us with a clear picture of biology as a subject matter. Pliny the Elder (23–79 CE) wrote a 37-volume summary of contemporary knowledge about the natural world, which became an authoritative text in Western Europe throughout the Middle Ages. Its title, *Natural History*, became the name for all inquiry into the properties of the observable world. At the start of book twelve, he clearly equates life with possession of a soul, in plants as well as animals and humans. Vegetable souls, or vegetable natures, reflect an important category in the observable world. All living things can be defined by a coherence that transcends the form present in rocks and other inanimate objects. All living things exhibit dynamic activities fueled by nutrition and maintained within physical bodies. Vegetable souls unambiguously label this property. Rather than inviting a dualist interpretation, they explain a perceived continuity. That would change in the Middle Ages as scholars came to associate souls, and particularly rational souls, with spiritual and subjective life-concepts.

References

Epicurus. *The Essential Epicurus*. Translated by Eugene O'Conner. New York: Prometheus Books, 1993.

Galen. *On the Natural Faculties*. Loeb Classical Library. Cambridge, MA: Harvard University Press, 1963.

Gerson, Lloyd. "Plotinus." In *Stanford Encyclopedia of Philosophy*, Summer 2014 ed. Stanford University, 1997–. https://plato.stanford.edu/archives/sum2014/entries/plotinus/.

Gotthelf, A. "Aristotle's Conception of Final Causality." *Review of Metaphysics* 30 (1976): 226–254.

Gundert, Beate. "Soma and Psyche in Hippocratic Medicine." In *Psyche and Soma: Physicians and Metaphysicians on the Mind-Body Problem from Antiquity to Enlightenment*, edited by John P. Wright and Paul Potter, 13–35 Oxford: Clarendon, 2000.

Hall, Matthew. *Plants as Persons: A Philosophical Botany*. Albany, NY: SUNY Press, 2011.

Johnson, Monte Ransome. *Aristotle on Teleology*. New York: Oxford University Press, 2005.

Lucretius. *The Nature of Things*. Translated by William Leonard. Boston: E. P. Dutton, 1922.

Plotinus. *Plotini Opera*. Edited by Paul Henry and Hans-Rudolph Schwyzer. Leiden, 1952a.

Plotinus. *The Six Enneads*. Translated by Stephen McKenna and B.S. Page. Encyclopedia Britannica, 1952b.

Singer, P.N. "Galen." In *Stanford Encyclopedia of Philosophy*, Winter 2016 ed. Stanford University, 1997–. https://plato.stanford.edu/archives/win2016/entries/galen/.

von Staden, Heinrich. "Body, Soul, and Nerves: Epicurus, Herophilus, Erasistratus, the Stoics, and Galen." In *Psyche and Soma: Physicians and Metaphysicians on the Mind-Body Problem from Antiquity to Enlightenment*, edited by John P. Wright and Paul Potter, 79–116. Oxford: Clarendon, 2000.

PART II

Development

CHAPTER 7

The Breath of Life: *Nephesh* in Hebrew Scriptures

Life-concepts have evolved over the centuries. Theories proliferate, divide, and fuse forming new species of meaning. The rise of elective monotheisms—Judaism, Christianity, and Islam—shaped thinking about souls and life. From the first-century onward, theologians, philosophers, and physicians worked to reconcile the classical three-soul typology with ideas about God and resurrection. Souls still covered the basic functions of biology, but they also took on new overtones.

Antique philosophers cared about what persists through change, generally, but Christians introduced a new challenge. They asked what persists through death. Two new life-concepts arose, neither of which mapped unambiguously to vegetable, animal, or rational life. Spiritual life refers to participation in the Divine life of God. The Hebrew book of Genesis describes God breathing life into the dust and forming a living soul (*nephesh*). Christian and Muslim theologians textured this with theories of resurrection life that included vegetable, animal, and rational activities. Internal life refers to subjective experience in sensation, reasoning, and will. Augustine introduced a new way of speaking about the interior life of humans. This required a realm of personal experience, reserved to the individual.

This reimagining of humans as spiritual and subjective opened a rift between vegetable and rational functions in humans, leaving animal functions in a strange limbo between them. Vegetable, animal, and rational souls, therefore drifted into a new configuration. This chapter looks at early Jewish life-concepts. Part II continues with the

© The Author(s) 2018
L. J. Mix, *Life Concepts from Aristotle to Darwin*,
https://doi.org/10.1007/978-3-319-96047-0_7

development of vegetable and animal souls in medieval theology and philosophy as various thinkers try to reconcile classical philosophy with scripture.

HELLENISTIC AND HEBREW THOUGHT

While Hellenistic thought moved from Plato to Plotinus, Hebrew thinkers wrote and reflected on life in their own way. They articulated no specific theory of plant life but contributed heavily to the conversation in three ways. First, they turned to scripture as a source of knowledge. Parallel origin stories led to a theory of dual creation with a profound impact on biology. Second, they introduced a new concept of soul and breath that reshaped the way both were used. Third, they opened the door to resurrection. This required rethinking life as well as death.

Under Macedonian and Roman rule, Hebrew culture was constantly in dialogue with Greek philosophy. By the beginning of the common era, many Jewish thinkers were reading their scripture in Greek and using Greek philosophical terminology. The best example appears in Philo of Alexandria, who tried to reconcile scripture with Plato. Jewish authors also started, very slowly, to consider ideas of immortal or immaterial souls.

As we turn to theology, it is particularly important to keep in mind symbolism and synecdoche. The terms "life" and "death" were used allegorically for good and evil, while terms like "body," "soul," "heart," and "mind" were used as stand-ins for the whole body or the whole self. This was true for Greek thinkers as well but, while the physicians and natural philosophers tended toward systematic precision, theologians and mystics used more narrative devices. It is important to read critically with an eye toward whether a simple reading was intended or something more suggestive.

HEBREW SCRIPTURES

Jewish and Christian philosophy has been heavily influenced by a collection of Hebrew texts commonly referred to as the *Tanakh* or Old Testament. The opening chapters of Genesis were among the most influential with regard to life-concepts. They came together around the sixth-century BCE but reflect a much older oral tradition. Genesis 1:1–2:3 includes the familiar six-day account of creation. God made the light (day 1), sea and sky/heavens (day 2), land and plants (day 3), stars and planets (day 4), sea and sky animals (day 5), and land animals including humans (day 6). On the seventh day God rested. God's role in creating all things and calling them good distinguishes Genesis from the perfect and imperfect creations in Plato's *Timaeus* and from the warring gods of Babylonian creation accounts. In Genesis 1, both plants and animals must be, in some sense, essentially in harmony with God and the universe.

A second creation account explains conflict within creation. Genesis 2:4–3:24 describes Adam and Eve, the first humans, and their disobedience. By eating forbidden fruit from the tree of the knowledge of good and evil, they broke the harmony of the world. This event is usually referred to as the Fall. Theologians (both Jewish and Christian) disagree on the extent to which this corrupted the essential character of all living things, but Genesis uses it to explain death as well as interspecies competition (Genesis 3:4, 14, and 18). Significantly, the sequence of events is not the same in the two accounts. Humans are created near the beginning of Genesis 2–3, but at the end of Genesis 1. Nor are these the only accounts of Creation in the Bible, though they are easily the most important for this discussion.[1]

In Roman times, scholars accessed these texts through their most prominent Greek translation, the Septuagint. That translation, dated to the third-century BCE, shaped later opinion through its vocabulary. Most interesting, the translators chose to equate the Hebrew word *nephesh* with the Greek word *psyche*. John Cooper provides this summary.

> An unbiased analysis of the biblical text itself, especially the Old Testament, will reveal that 'soul' and 'spirit' (Hebrew: *nephesh* and *ruach*; Greek: *psyche* and *pneuma*) have quite different meanings than they do in Platonic circles. They are used of animals and humans alike and have more to do with the power of life and breath in the earthly creature than anything remotely connected with immortal existence after death. (Cooper 1989, p. 3)

The primary life distinction made in Hebrew scriptures centers on the word *nephesh*, commonly translated as 'living being' or 'soul.' In

---

[1] Brown (2010) lists seven: Genesis 1:1–2:3, Genesis 2:4–3:24, Job 38–41, Psalm 104, Proverbs 8:22–31, Ecclesiastes, and (Second) Isaiah.

Genesis 1:20–24, God causes living beings (*nephesh chay*) to arise from sky, sea, and earth. Similarly, in Genesis 2:7, when God breathes into the dust, Adam becomes a living being. Throughout the Hebrew scriptures, humans and other animals are spoken of as creatures with breath. God's breath (*ruach* or *neshamah*) stirs up the dust, changing nonliving, tangible material into a dynamic, relational organism (e.g., Ezekiel 37:5–14). Similarly, when the breath departs, the organism falls again to dust (e.g., Job 34:14–15). Cooper (1989, pp. 45–51) compares this to Aristotle's hylomorphism: the Hebrew living being is a single whole, requiring both matter and dynamic breath. It is not dualist, but holist. To be a living being is to be the kind of thing that has breath moving in it. Joel Green (2008) takes this one step further, claiming that the Hebrew concept of soul is not essentialist, but fundamentally relational; animals are made to be in relationship with other animals and with their environment.

Living being always have corporeal bodies. As with Plato, Aristotle, and their predecessors, Hebrew scriptures describe integrated physiological and psychological functions. The same terminology applies to both. They locate livingness and soul in the breath and in the blood, suggesting parallels with the refined *pneuma* of the Greek physicians (e.g., Genesis 9:4; Leviticus 17:11). They explicitly tie this breath to Divine action. Greek talk of pneuma emphasized the physical and elemental aspects of life-concepts; mind was associated with transcendence. Hebrew terms, on the other hand, suggest an intimate connection between breath and God's action in the world. Psychological identity was placed, literally and figuratively, in the heart and the bowels.

Plants, meanwhile, were treated as part of the landscape, created for the convenience of living beings. Mark Smith (2010) suggests that Genesis 1 should be read as a story in two acts.[2] In the first act (days 1–3), God made the regions of the world—first the heavens, then the sky and sea, and finally the land. In the second act (days 4–6), God filled the regions with citizens; stars and planets inhabit the heavens, birds and fish inhabit the sky and sea, and finally animals (including humans) fill the land. William Brown (2010, pp. 36–44) compares this to the precincts of the Temple, suggesting a procession from the profane outer courts into the Holy of Holies. The regions and inhabitants are introduced in order

---

[2] Aquinas also makes this argument in *Summa Theologiae* 1.70–72.

of ascending significance in a way that points toward the throne of God, signified by the seventh day, the Sabbath. Plants are furniture; their value dependent more on their location than their action.

Little attention was given to plant life, per se. It exists for the sake of feeding and sheltering animal life.[3] Over the last two millennia, many theologians have discussed the role of humans in relation to other animals, often questioning human superiority. Hierarchical views—making humans the end and epitome of creation—fit well with Genesis 1, but face critiques even within scripture with the Fall of humanity in Genesis 2–3 and life's futility in the wisdom literature (Mix 2016; Gustafson 1994). Nonetheless, plants remain in the background, only used figuratively to express abundant or fleeting life (e.g., Psalm 103.15; Isaiah 40:6–8).

Animal life was always associated with a physical body, earthly processes, and finite time (Brown 2014). Animals, including humans, do not possess a soul (*nephesh chay*); they are a soul, and that only while the breath moves the dust. Hebrew authors used an entirely different word, 'shade' (*rephaim*) to refer to the remnant that persists after death (Cooper 1989, pp. 52–62). Scholars argue on how and when doctrines of immaterial souls and resurrection arose within Israelite religion. With a few debated exceptions, life in the Hebrew scriptures always appears in a material, physical context.

A Second Life

Centuries of speculation about life after death looked for hints of resurrection in the Hebrew scriptures. The integrated breath and dust of Hebrew life covers both physiological and psychological traits within animals during life. After life, however, the connection was less clear. Cooper (1989) and others argue for a nascent dualism within scriptural resurrection accounts. This has become the dominant narrative in most histories of the soul (e.g., Martin and Barresi 2006, pp. 39–44; Goetz and Taliaferro 2011, p. 31). A closer look at the passages in question suggests that allegorical interpretations are equally plausible, especially for the earlier texts. The underworld of Sheol appears, much like Hades,

---

[3]For a summary, see Hall (2011, pp. 57–66). Note that plants are given for the benefit of all animals, not just humans, in Genesis 1:29–30. Humans are not specifically given permission to eat other animals until the Noahic covenant at Genesis 9:2–3.

as a repository for fading phantoms, not a waiting place (e.g., Genesis 37:35, Psalm 6:5). Ecclesiastes says, "Whatever it is in your power to do, do with all your might. For there is no action, no reasoning, no learning, no wisdom in Sheol, where you are going" (Ecclesiastes 9:10, JPS Translation; see also 3:19–21 and 12:7). Several passages speak of a return from death (i.e., Isaiah 26:19, Ezekiel 37:1–14, Hosea 6:2). These resurrections, however, occur locally (only one or a few are raised) and give no indication that the new life is any less physical or mortal than normal human life. It seems plausible that these passages, like the resurrection miracles of Elijah and Elisha, were meant to be taken allegorically, to display God's power, and prophetically, to promise the restoration of Israel (I Kings 17:22, 2 Kings 4:35, 13:21). Two passages stand out. Job speaks of seeing redemption in the flesh (Job 19:26–27) and Daniel says that some who sleep (in death) will rise (Daniel 12:2). Likely composed later than other parts of the *Tanakh*, they are more amenable to a literal interpretation, though they could be allegorical as well. The only clear example of new or continuing life along Platonic lines appears in the Wisdom of Solomon, an apocryphal book written in the first-century (BCE or CE) and showing strong Greek influences (2:23–24; see Coogan 2001, p. 70 for commentary). Here, humans were made incorruptible, but through the envy of the Devil, became mortal.

After the consolidation of the *Tanakh* as a single volume and before the writing of the Christian scriptures (first-and second-centuries CE), doctrines of resurrection slowly evolved, pushing the human soul farther away from the dynamic processes of the mortal body. The three main camps that arise within Israelite culture favor soul annihilation, soul reincarnation, and body/soul dualism.

One camp claimed human life ends with the body. Numerous scriptural passages support this position, often in comparison with the life of plants. In the book of Job (14:7–12) human life is presented as more finite than plant life.

> For there is hope for a tree, if it is cut down, that it will sprout again, and that its shoots will not cease. Though its root grows old in the earth, and its stump dies in the ground, yet at the scent of water it will bud and put forth branches like a young plant. But mortals die, and are laid low; humans expire, and where are they?[4]

---

[4]See also Isaiah 26:14 and Ecclesiastes 3:19.

Many other passages compare the transience of human life to the brief lifetime of grass. Both Flavius Josephus and the Christian Gospels gave this position to the Sadducees.[5] Named for Zadok, the first High Priest in Solomon's Temple, they were the establishment party within Israelite religion.

A second camp spoke of the soul returning in a reconstituted or entirely new body. This resurrection is both material and physical, though the rules may shift in the transition. The return of soul to flesh, dust, or earth appears in Ezekiel 37:5–14, Job 19:26–27, and Isaiah 26:19. Alternatively, traditions developed suggesting that the new body would be made of more refined materials, such as air, fire, or aether. Daniel 12:3 hints in this direction, but the idea appears more clearly in the apocryphal works of II Esdras (75–101) and II Baruch (51). These more refined, more spiritual bodies would be incorruptible, or less corruptible, than the gross bodies of mortal life. The New Testament attributes fleshly resurrection to the Pharisees, a group of Israelites favoring strict piety (Acts 23:6–7). Josephus also makes this connection.[6]

Arguments have been made for a third camp, promoting dualism, with a spirit or soul leaving the body and persisting on its own. Such a position may be suggested by Ecclesiastes 3:21: "Who knows whether the human spirit goes upwards and the spirit of animals goes downwards to the earth?" An Israelite sect called the Essenes advocated for an immortal and immaterial soul made of refined air.[7]

Hebrew scriptures provide little and controversial evidence for doctrines of resurrection life. The evidence they do provide suggests that it followed the same rules as daily, mortal life in the flesh. The resurrection did not provide a new life-concept; rather, it spoke to the power of God working in the world. Nonetheless, Israelite culture and religion showed clear signs of a growing interest in immortal and immaterial life-concepts. Those ideas opened a space for a new and different life that would be characterized as spiritual and eternal.

---

[5] Josephus (1984: Vol. IV, pp. 3–4)—Antiquities of the Jews VIII.1.4. In the New Testament, see Matthew 22:23, Mark 12:18, Luke 20:27, Acts 4:1–2, and Acts 23:8.

[6] Josephus speaks of incorruptible souls, eternal punishment of the wicked, and reincarnation of the just, perhaps following Plato. Josephus (1984: Vol. I, pp. 150–151) (Wars of the Jews II.8.14); see also Vol. IV, p. 3 (Antiquities VIII.1.3).

[7] Josephus, "Wars of the Jews" II.8.11 in Vol. I, p. 149.

PHILO OF ALEXANDRIA

Philo presented the first systematic integration of Hebrew and Greek life-concepts. A Hellenistic Jew living in Alexandria (c. 20 BCE–c. 40 CE) he attempted to show that Platonic and Stoic cosmology were entirely consistent scripture. Philo located body and soul within a Platonic continuum, but his dual-creation history and ontology formed the nucleus for a radical divide that grew throughout the Middle Ages.

Philo used Plato's *Timaeus* and Genesis to form one coherent picture of creation, which he presented in *On the Account of the World's Creation Given by Moses* (Philo 1949: *On Creation*). He divided the creation into two events. Through the Divine reason, God created the intelligible forms of things. This invisible creation or "creation in light" corresponds to Genesis 1:1-3 and to the work of the Creator in *Timaeus*. God then created the corporeal objects which exhibit the forms. The visible creation corresponds to the work of the gods in *Timaeus* and to the six days of creation in the remainder of Genesis 1. Speaking of time in this scheme is somewhat awkward. The ideal, invisible creation occurs eternally while the material, visible creation played out after time was created on the first day. Thus, the invisible creation is ontologically and explanatorily prior, but not chronologically first. Like Plotinus, Philo thought in terms of a cosmic intellect acting outside the course of normal time.

The divided cosmos lends itself to dualism, but still represents the dynamic relationships of Plato. Regarding humans and plants, the division is clear. One participates in the invisible creation in a way the other does not. Looking at other animals, however, blurs the picture. Philo drew an analogy between soul and body. The soul perceives incorporeal reality by way of the intellect as the body perceives corporeal reality by way of the eye (sec. 17). Physical light represents ideal knowledge. Humans, with a rational soul and intellect, participate in the invisible creation; they perceive true knowledge. Other animals, with a sensitive soul and sight, participate by analogy; they perceive the physical world. Plants, lacking both, are mentally inert. This is not to say that plants do not participate in some way. They were created fully formed in the visible creation and they perpetuate their type through reproduction. Thus, they were present in the invisible creation in form and principle. Still, once created, they cannot interact with the invisible in any other way.

Philo associated the soul (*psyche*) and living things (*zoa*) with both sensation and locomotion, drawing the now familiar line between animals and non-animals. He then blurs the line with two curious exceptions: stars and fish. Stars adorn the heavens, just as plants adorn the land.[8] The plants are given as food for animals and the stars as food for thought; their rhythm inspires philosophy. One might infer that stars are similarly disregarded as cosmic furniture, but Philo speaks of them as rational animals. They move and perceive the ideal world. This makes them intermediate between humans and plants. Philo suggested that God created the plants before the stars to remind us that stars should not be worshiped.

Philo referred to humans as the most favored living things (secs. 24–25). God willed that they not only live, but live well, and so granted them a full banquet of both corporeal and ideal stimuli. They perceive the material world with senses and the ideal world with the intellect. Stars lack corporeal perception, just as the irrational animals lack ideal perception. Thus, both are less than human, but more than plants.

Fish, on the other hand, participate in the physical world, but are only nominally ensouled (sec. 22). God made them first among the animals and granted them nutrition, reproduction, breath, and movement, but not sensation. Thus, they are animals and not animals, soulless movers. Their bodies overwhelm their souls. This provides a two-dimensional continuum of ensoulment. Among corporeal things, the axis runs from the barely ensouled, plant-like fish to the fully living humans. Philo recognizes a number of rational activities in animals (notably bees) and even plants, but attributes it to Divine intelligence working through them rather than any inherent intellect (*De Animalibus* 92–95, in Terian 1981, pp. 106–107). Among ideal things, both humans and stars have soul and mind, but only humans are subject to corporeal influence.

Philo read the second creation story in Genesis allegorically, interpreting it as an account of mind and reason (Philo 1949: *Allegorical Interpretation*). Ironically, the allegorical account gave him space to speak more concretely about plants and animals. He reiterated the fourfold hierarchy of the Stoics, with inanimate coherence, plant growth, animal sensation, and rational intellect (sec. 7). He used Plato's anthropology with three aspects of the human soul present in three parts of the

---

[8] *On Creation* sec. 12–19. Philo suggests that the statement "Let us make humankind" (Genesis 1:26) means that God collaborated with created subordinates when making humanity. Compare the collaboration between the Creator and the gods in *Timaeus*. This co-creation made humans susceptible to vice.

body: reason in the head, spirit in the chest, and physical desire in the abdomen (sec. 22). Finally, there is a fascinating return to corporeal soul theory, in which he spoke of the human body being made of earth and the human soul being made of rarified air (sec. 55).

One can also see the beginnings of an analogy popular throughout the Middle Ages. Man is to woman as mind is to body. In exploring Genesis 2 as an allegory for human composition, Philo spoke of the creation of Adam as the birth of mind and intellect, the creation of Eve as the birth of physical sensation (Philo 1961, p. 30).

In Philo, the soul continued to play two roles. Still intimately tied to nutrition and reproduction among animals, it has been severed from the plants. Among the animals, it relates to sensation, locomotion, and reason, but need not be associated with all three. None of these souls is necessarily immortal, though the rational soul may become so through interaction with Divine reason. Philo's thought provided a strong influence for Augustine and other Christians but enjoyed far less prominence among Jewish thinkers.

The distinction between religious and secular reasoning can be hard to track historically. The *Tanakh* was not the first document to insert questions of metaphysics, value, and ethics into the discussion. Nor was it the first document to claim some form of dogmatic or spiritual authority. Nonetheless, it introduced background stories, particularly in Genesis, that came to dominate discussions of biology in Europe. Until at least the sixteenth century, philosophers of biology felt obligated to tie their accounts to scriptural stories, even when they disagreed with them.

That effort required a recognition of the two creation stories and attendant questions about how literally they should be interpreted. The plain sense meaning of the first two creation stories are not compatible, leading to extended debate about how one should understand specific passages, including those on the creation of plants, animals, and humans. Until the Protestant Reformation, theologians almost universally accepted that figurative, allegorical, and analogical interpretations would be required, either in addition to or in place of plain sense readings. Once that door had been opened, all of scripture could be read in multiple ways, including resurrection accounts.

Baseline Hebrew theories of life from this period include two life-concepts. First, plants represent God's action, making the land habitable for animals. They act as furniture, providing food and shelter for other forms of life. Second, God's breath enlivens the dust to make *nephesh chay*, living souls, which include humans and all animals. Those souls are unmistakably physical. They can be known by the movement of blood and breath within a creature. Indeed, blood and breath both worked as synonyms for, if not explanations of, life. All life is corporeal. All life is mortal. Human shades persist in Sheol, but only as the fading echo of a once-living being.

Belief in an afterlife appeared slowly. Scriptural accounts emphasized the power of God and foretold the revival of Israel, after disobedience, destruction, and death. They tended toward simple resuscitation of physical, animal life, rather than introducing a new concept. Interaction with Hellenistic culture led to more Platonic ideas of immortal souls and transmigration by the start of the common era.

Philo of Alexandria synthesized Greek and Hebrew accounts. His dual creation Platonized "light" and divided bodies from forms. His writing about the *Logos* revealed a Platonic view of creation manifesting through mind and soul. Pharisaic resurrection and Philo's dual creation would both have a profound impact on later Christian life-concepts.

REFERENCES

Brown, William P. *The Seven Pillars of Creation: The Bible, Science, and the Ecology of Wonder.* New York: Oxford University Press, 2010.

Brown, William P. "Life." In *Oxford Encyclopedia of the Bible and Ethics*, edited by Robert L. Brawley. New York: Oxford University Press, 2014.

Coogan, Michael D., ed. *The New Oxford Annotated Bible with the Apocryphal/Deuterocanonical Books*, 3rd ed. New York: Oxford University Press, 2001.

Cooper, John W. *Body, Soul, and Life Everlasting.* Grand Rapids, MI: Eerdmans, 1989.

Goetz, Stewart, and Charles Taliaferro. *A Brief History of the Soul.* Malden, MA: Wiley-Blackwell, 2011.

Green, Joel B. *Body, Soul, and Human Life.* Grand Rapids, MI: Baker Academic, 2008.

Gustafson, James M. *Sense of the Divine: The Natural Environment from a Theocentric Perspective.* Cleveland: Pilgrim Press, 1994.

Hall, Matthew. *Plants as Persons: A Philosophical Botany.* Albany, NY: SUNY Press, 2011.

Jewish Publication Society (JPS). *Tanakh, the Holy Scriptures.* New York: Jewish Publication Society, 1985.

Josephus, Flavius. *The Works of Josephus*, 4 vols. Translated by William Whiston. Grand Rapids, MI: Baker Book House, 1984.

Martin, Raymond, and John Barresi. *The Rise and Fall of Soul and Self: An Intellectual History of Personal Identity*. New York: Columbia University Press, 2006.

Mix, Lucas J. "Life-Value Narratives and the Impact of Astrobiology on Christian Ethics." *Zygon* 51 no. 2 (2016): 520–535.

Philo. *Philo*, 2 vols. *Loeb Classical Library*. Translated by F.H. Colson and G.H. Whitaker. Cambridge, MA: Harvard University Press, 1949.

Philo. *Philo Supplement I: Questions and Answers on Genesis. Loeb Classical Library*. Translated by R. Marcus. Cambridge, MA: Harvard University Press, 1961.

Smith, Mark S. *The Priestly Vision of Genesis 1*. Minneapolis: Fortress Press, 2010.

Terian, Abraham. *Philonis Alexandrini de Animalibus: The Armenian Text with an Introduction, Translation, and Commentary*. Ann Arbor, MI: Scholars Press, 1981.

CHAPTER 8

Life After Life: Spiritual Life in Christianity

Christian theology began to truly divide physiology and psychology with the introduction of a second life. In Jewish thinking, all life was an expression of Divine action in the world. The New Testament shows elements of this, but also speaks of a second birth and a second death. Spiritual life became associated with this new life and with resurrection, each of which had vegetable, animal, and rational elements. Trying to understand biology in light of both Greek souls and the New Testament, early Christian authors took two approaches. Tertullian embraced the materialist approach of the Stoics while Origen embraced the transcendent approach of Plato.

A Note on Interpretation

Christians have argued about the resurrection from the beginning of Christianity. To what extent should it be considered material and physical? To what extent does resurrection life continue along the same lines as mortal life? I will not resolve either the interpretation question (what the New Testament says) or the doctrinal question (what Christians should believe). For this history of life-concepts, I only wish to show that a new life-concept has arisen alongside vegetable, animal, and rational life and that aspects of it map onto all three.

© The Author(s) 2018
L. J. Mix, *Life Concepts from Aristotle to Darwin*,
https://doi.org/10.1007/978-3-319-96047-0_8

Two Lives, Two Deaths

The Christian New Testament contains 27 books composed in Greek in the late first and early second century of the common era. The earliest works, letters attributed to Paul of Tarsus, display a sophisticated philosophical vocabulary, but address specific moral questions. The Gospels, written later, focus on telling the story of Jesus' life. The New Testament also contains a number of other letters and the book of Revelation, a dramatic depiction of the end of the world, variously interpreted as prediction and social critique. The New Testament does not contain either an explicit natural philosophy or a creation story. With the plurality of life-concepts in the Hellenistic world, no consensus has been reached about how best to interpret words such as flesh (*sarx*), body (*soma*), soul (*psyche*), and spirit/breath (*pneuma*). Should they be read through the lens of Platonic idealism, Aristotelian activities, or Stoic and Epicurean materialism? Do they reflect early Hebrew holism or a later dualism?

SPIRIT VERSUS FLESH?

Joel Green (2008, pp. 51–61) provides useful cautions against naïve application of Cartesian dualism, Greek or Hebrew word equivalences, or even a clear divide between Greek and Hebrew thought. Rather, the text itself represents something new arising within a swelter of competing doctrines. A number of passages contrast spirit and flesh. Green argues that both John's Gospel and Paul's letters provide a "dialectical rather than an ontological division" (p. 11, see also Heckel 2000). In other words, both use the distinction between bodily life and spiritual life as a way of speaking about a unified whole whose spiritual aspect is more significant or closer to the core of human identity.

In the Gospels, Jesus compares physical healing and personal forgiveness in a way that invites comparison and may suggest that the two are integrated (Matthew 9:2–8, Mark 2:3–11, Luke 5:17–26). Physiological, psychological, and spiritual health were integrated for both Greek and Hebrew thinkers. Both flesh and spirit shape us, body and soul. Their interaction makes the comparison relevant. If flesh and spirit were truly independent, bodily health could not impact spiritual life (Martin 1995, p. 174).

The word *pneuma* has spiritual and psychological connotations. Indeed, spirit is the most common translation. It also has bodily and physiological connotations. This appears most dramatically in the death of Jesus, uniformly described as a loss of breath. Jesus surrenders his pneuma to God and breathes for the last time.[1] The literal expiration of Jesus marks a bodily change and a spiritual change. Thus, we should not jump too quickly to assume that other references to breath are necessarily immaterial or nonphysical. Corporeal breath was still tightly linked to spiritual health. Numerous scriptural commentaries cover the question of the human soul (see summaries in Green 2008; Murphy 2006; Cooper 1989). Far more important for our purposes will be the language of two lives (and two deaths) present throughout the New Testament.

Two Deaths

Resurrection was a late-comer in Hebrew thought; it proved central to Christianity. The New Testament contains two distinct words and two distinct concepts that were, unfortunately, conflated in the King James Bible and other translations: Hades (αἴδης) and Gehenna (γέεννα). Both can be translated as Hell, but the Gospels of Matthew and Luke, as well as many later Christian authors, viewed them as distinct.

Hades is a translation of the Hebrew *Sheol*.[2] In Revelation, death and Hades are presented as temporary soul repositories, which will be emptied, judged, and cast into the fire (1:18, 6:8, 20:13, 20:14). They represent the end of physiological life. Insofar as there is eternal punishment or reward, it occurs after Hades.[3]

---

[1] Matthew 27:50, Mark 15:37, Luke 23:46, John 19:30. The physiological meaning is retained in the NRSV translation "breathed his last" while the King James Version sounds less physical, "gave up the ghost." Curiously the death of Jesus is marked by loss of breath (*ekpneo*) in Mark and Luke, while the death of Ananias (Acts 5:5), Sapphira (Acts 5:10), and Herod (Acts 12:23) is marked by a loss of soul (*ekpsycho*). Neither word appears elsewhere in the New Testament. Likely, both represent colloquial expressions for death.

[2] Acts 2:25–28 quoting Psalm 16:8–11; I Corinthians 15:55 quoting Hosea 13:14.

[3] One remaining account occurs in the Parable of Lazarus and the Rich Man. Jesus speaks in Luke (16:19–31) about a rich man punished for neglecting the poor when he was alive. Debate persists among Biblical scholars about how literally this story should be taken. Is it a report on the afterlife or simply a parable? Many Christian commentators have used it to develop doctrines of punishment and reward while waiting for Jesus to return. See Cooper (1989, pp. 124–127).

Gehenna refers to a trash heap on the outskirts of Jerusalem, frequently with some sense of things discarded or cast into the fire. It appears in conjunction with judgment or condemnation, sometimes explicitly post-death. Most references simply invoke judgment (e.g., Matthew 23:15; Mark 9:43; Luke 12:4–5). Gehenna parallels the Greek Tartarus, the flaming land of eternal punishment below Hades.

If one can be recovered from death, then several questions arise. What goes to the land of the dead? What goes on to punishment or reward? The New Testament promises resurrection, suggesting that something persists through death and through Hades. From here on out, Christians would argue about what that is. It may be a personal, psychological soul or an impersonal spirit. Or, it may be that humans cease to be at death and are entirely recreated, body and soul, in the new creation. All three possibilities were explored in the following centuries.

Up to this point, souls were either eternal, but experienced in physiological bodies (as in Pythagoras, Plato, and Plotinus), or corporeal processes that explain physiology and psychology (as in Aristotle and the Stoics). There may have been souls without bodies, but there were never living bodies without souls.[4] The Pharisees and Paul both argued for resurrected, spiritual bodies, but the earliest Christians were already talking about immaterial resurrection. More significantly, the New Testament suggests that living bodies (with souls) might lack spiritual life.

Two Births

The Gospel of John proposes a second birth. "But to all who received him, who believed in his name, he gave power to become children of God, who were born, not of blood or of the will of the flesh or of the will of man, but of God" (John 1:12–13).[5] Jesus reaffirms this later when he tells Nicodemus, "Very truly, I tell you, no one can enter the kingdom of God without being born of water and Spirit. What is born

---

[4]To be precise, there were never living bodies without souls or some equivalent organizing entity such as a pneuma *physikon*. Even Philo placed vegetable souls within the invisible creation.

[5]Note that being born "of blood" reflects a physiological life-concept (cp. vegetable soul), being born "of the will of the flesh" reflects a sensitive/agential life-concept (cp. animal soul), and being born "of the will of man" reflects a human-specific or rational life-concept (cp. rational soul). This new birth is none of the three.

of the flesh is flesh, and what is born of the Spirit is spirit" (John 3:5–6). A new life-concept has been introduced. This language of rebirth occurs throughout the New Testament. I will refer to this as spiritual life from here on out. Significantly, spiritual life is not identified with a rational function, because rational humans (e.g., Nicodemus) might not have it.

First Corinthians (15:44) describes the resurrection, thus. "It is sown a physical body, it is raised a spiritual body. If there is a physical body, there is also a spiritual body."[6] This translation (New Revised Standard Version) suggests that Paul is speaking of a material, physical entity in the first instance. The Greek text for "physical body," however, is *soma psychikon*, the soul-oriented body. Paul has a concept of mortal life as soul-oriented and opposes it to the *soma pneumatikon*, the breath/spirit-oriented body of the resurrection. Thus, the spiritual life cannot be equated with the soul, either.

Dale Martin (1995, pp. 117–120) suggests that the new spiritual life could be identified with a rarified body, like that of stars in Plato, Baruch, and Philo. First Corinthians (15:39–47) hints that in the resurrection, human bodies will be like the bodies of stars and planets. Alternatively, the Holy Spirit (as *logos* or Divine *pneuma*) may direct the human *psyche* and *pneuma* through the mind (Martin 1995, p. 98). Theo Heckel (2000, p. 125) argues that Paul was reaching for a temporal rather than a spatial or substantial distinction. Before Jesus and baptism, humans are focused on self; afterwards, they participate in something larger than themselves. The church as the "body of Christ" becomes an organism, whose breath integrates even the soul and mind of individual Christians into something greater.

In Genesis and throughout the Hebrew scriptures, the divine breath enlivens all animals physiologically. The New Testament changes this. Whether by opposing mortal and Divine breath or distinguishing between types of Divine breath, a space has opened between two life-concepts: mortal and spiritual. Mortal life includes vegetable, animal, and rational activities; spiritual life may as well, though only in the context of Divine life. The text is unclear where souls fit in this scheme. Spiritual life begins in baptism, when the Holy Spirit "enlivens" an ensouled, bodily human life.

---

[6]For commentary, see Green (2008, pp. 172–175), Martin (1995, pp. 63, 123–129), and Cooper (1989, pp. 139–141).

The opposite of spiritual life was not physical life, but spiritual death. Thus, Gehenna might be read as mortality in spirit, the complete death. This appears most clearly at Matthew 10:28: "Do not fear those who kill the body but cannot kill the soul; rather fear him who can destroy both soul and body in [Gehenna]."

CHRISTIAN IDENTITY

Personal continuity took on a new urgency with the New Testament. Something still persists through life, providing biological continuity. Something must also persist through death, providing personal continuity. Are they the same? If not, how are they related? The problem of unity and multiplicity returned in a new form.

A Platonic continuum from base matter to ideal order would not work in Christianity, because God made humans, body and soul, and found them good. God promised resurrection in the flesh. Worse yet, sin and death did not arise from materiality, but from a desire to be like God, and a choice for body over spirit (Heckel 2000, pp. 123–124). For Christians, evil came not from materiality and vegetable desires, but from the will.

Tertullian

Tertullian (c. 155–c. 220) provided the prototype for material soul-concepts in early Christianity. Growing up in Carthage, he converted to Christianity around age twenty, but never felt truly satisfied with orthodox positions. He first joined the Montanist sect and then founded his own group. Throughout, he made a name for himself as a philosopher and apologist. Drawing on Epicureans, Stoics, and physicians, he championed corporeal souls. They provided continuity between parents and offspring and between mortal and spiritual life.

Tertullian wrote his own book, *On the Soul* (Tertullian 1885, 1947). Like Lucretius, he used the Latin terms *anima* and *animus*. The first more closely aligns with the Greek *psyche*, encompassing all life-concepts from nutrition to reason. The second, he compared to mind. Citing Plato's threefold division, he argued that these are neither separate souls nor even aspects of the same soul so much as functions carried out by the soul, which is, in itself, a substance (secs. 14–16). As such, there is but one soul, which is best identified with the rational function. It rules both mind and body. Unlike Plato and Aristotle, Tertullian was anxious to keep the senses and the intellect together (sec. 18). Bodily sensation and mental intellection give us access to different things (bodies and ideas) but the same soul perceives both. He believed that separating them leads to Gnostic dualism. He also argued that the human soul and the human spirit are the same thing (sec. 10).[7] The physical function of breathing and the spiritual function of being moved by the breath of God are both essential to the soul's nature. In integrating body and soul, Tertullian stated that both come into being at the same time and both come from the substance of the parents, a doctrine that came to be known as *traducianism* (sec. 27).

This willingness to wrap all aspects of the soul into a single, material entity gave Tertullian a language of the soul that is unitary. Humans have one life: the soul. It is both corporeal and immortal, though it enlivens a grosser corporeal body that can die. Human mind occurs naturally in the context of animal soul and all soul occurs naturally in the context of vegetable life. Citing the Stoics and arguing against Plato, he said that the soul must be corporeal because it is affected by the physical world and suffers pain in Hades, even without a grosser body (secs. 5–7). When Jesus comes again, souls will be rejoined to bodies (miraculously the same bodies) and judged. The good pass on to Paradise where their bodies will be free from corruption (sec. 55). The evil do not live forever but die forever, body and soul, in Gehenna (*On the Resurrection of the Flesh*, 35). Similar ideas can be found in Irenaeus of Lyon (mid-second century) and Marcus Minucius Felix (early third century) (Martin and Barresi 2006, pp. 58–61).

Only humans have "souls" in Tertullian, but he makes it clear that animals and plants have parallel life-drivers. The important distinction being that they are not in the image of God. Curiously, Tertullian remarks that plants are not limited to appetitive function. The creeping vine demonstrates that plants can have sensation, knowledge, and even intellect (*On the Soul*, 19). They do not have spiritual life.

---

[7] See also *Against Marcion* 2.9 where he argued that the soul cannot be the same as the Spirit of God, because the soul can err and is a life while the Spirit is life-giving. The human spirit resembles the spirited or animal soul, while the Spirit of God enlivens the mind or rational soul.

Origen

Born in Alexandria, Origen (c. 185–c. 254) traveled throughout his lifetime. One of the earliest systematic thinkers in the Christian tradition, he incorporated Greek philosophy and New Testament theology. He was considered a heretic from at least the sixth century for ideas too close to Neoplatonism and Gnosticism, but his thought influenced theology in both Eastern and Western Christianity.

Origen's basic physics, biology, and cosmology can be found in his treatise *On First Principles* (Origen 1966). He subscribed to a form of hylomorphism, claiming that no entities other than God can exist without a body (2.2–3). This includes souls. Following, Aristotle, Origen saw the defining characteristic of life as self-motion; all life has a cause of motion characteristic of and intrinsic to itself (3.1.2). This dynamism warrants the label "soul" and, thus, he can speak of the souls of plants, planets, and angels, each a material, physical entity responsible for movement.[8] Despite mentioning this traditional view of vegetable souls, he used different concepts at different times, highlighting vegetable, animal, rational, and spiritual components in different works. A second definition, identifying souls with sensation and locomotion, matches more closely with Aristotle's sensitive soul. He specifically argues that this soul is present in the blood of all animals, including fish and insects (2.8.1). Elsewhere, he spoke of the soul as applying only to rational beings.

Origen reasoned from the ethereal bodies of stars and planets (made of rarified air or fire) to ethereal bodies in the resurrection (2.3.2).[9] They do not suffer corruption the way that gross, earthly bodies do, but they are nonetheless material and physical. The human soul shifts from an earthly body to a spiritual one after death. He even claims that the soul, when apart from earthly or ethereal host, must be corporeal in some sense if it is to be visible. One cannot see a ghost unless it has a body of some sort (*Against Celsus*, 2.60, Origen 1979).

---

[8] Plants: *On First Principles* 3.1.2, see also *Against Celsus* 4.58, 74, 83; Planets: *On First Principles* 1.7; Angels: *On First Principles* 2.8.2, though see also speculations at *Against Celsus* 3.37.

[9] The position appears to be taken literally here, but is used to set up an allegorical argument, drawing on Genesis 9:4 and Leviticus 17:11, about the inward man in *Dialogue with Heracleides*.

Just as the soul can be corporeal, Origen also believed that the body has a dynamic form that defines it. In defending resurrection life within an identical body, he compared the form of the body to a river.

> Because each body is held together by [virtue of] a nature that assimilates into itself from without certain things for nourishment and, corresponding to the things added, excretes other things..., the material substratum is never the same. For this reason, river is not a bad name for the body since, strictly speaking, the initial substratum in our bodies is perhaps not the same for even two days.
>
> Yet the real Paul or Peter, so to speak, is always the same—[and] not merely in [the] soul, whose substance neither flows through us nor has anything ever added [to it]—even if the nature of the body is in a state of flux, because the form (*eidos*) characterizing the body is the same, just as the features constituting the corporeal quality of Peter and Paul remain the same. (Bynum 1995, p. 64 quoting a fragment of Origen's commentary on Psalm 1)

We might call this the vegetable soul, though given the text, it may be more precise to call it a living nature. Caroline Bynum (1995, pp. 63–68), in citing this fragment of Origen, argues that the river metaphor owes more to Platonic and Stoic explanation (as an internal motivator of development) than to Aristotelian concepts of nutrition. While it does not resemble the later entelechy of Thomistic "Aristotelianism," it does fit well with the nutritive soul of Aristotle. The path and flow of the river align with the nutritive soul as formal cause of life in action and fulfillment. In any case, the quote demonstrates a dynamic bodily form within Origen. Bynum even compares the agency of such a bodily form to the agency of genes in biological development. This process-oriented idea of life and resurrection was later taken up by Gregory of Nyssa (335–395 CE).[10]

Origen developed an idea that comes to be called trichotomism, the doctrine that humans are composed of three parts: body, soul, and spirit. For Origen, the spirit was perfect and hot in its perfection. Souls represent spirit that has cooled in its fervor and love of God (*On First*

---

[10]Martin and Barresi 65–68; see further in Gregory's (1993) *The Soul and Resurrection*, where he both named and endorsed the vegetable soul as the appetitive agency in Plato and nutritive function in Aristotle. In *On the Making of Man*, 8.4, he also set out the three Aristotelian souls before saying that the soul most properly refers to the rational soul.

Principles 2.8.3). The soul, in turn, organizes the body. In his first "Homily on Genesis," Origen (1982) compared the spirit, soul, and body to a love triangle. The human soul is like a woman who is married to the spirit. She may find duty and joy in her spouse or have an affair with the body. This three-level anthropology, along with the value orientation toward a higher, spiritual life echoes the Platonic three-part soul. Hunger and bodily desire are specifically associated with the vegetable life-concept identified here as body. It remains unclear whether the human reason and will properly belong with the soul or the spirit. In either case, they can be blinkered by the body.[11]

None of Origen's extant works develop this further into a dualism of matter and spirit, though that elaboration seemed unavoidable to many later commentators (Edwards 2014). Origen (1982, p. 48) briefly mentioned a twofold creation along the lines of Philo. He elaborated that idea in reflections on the Trinity, explicitly linking Christ with the *logos* of the cosmos (*Against Celsus*, 2.31).

In the preface of *On First Principles*, Origen claimed agnosticism about whether souls are preformed or produced from pieces of the souls of parents (traducianism). Nonetheless, later theologians associated the following account with Origen. God created all rational souls in the invisible creation and held them in a repository. Prior to the beginning of time, many of the souls rebelled, turning from God. They cooled and became more fleshly, more mortal, so that those who drifted the least became angels, then humans, then demons. The history of the cosmos is defined as God's gradual process of bringing all those souls back into a state of perfect joy.

In Origen's picture of the world, we have Jesus Christ, acting as the power above and behind all powers, bringing the world into being. Drawing on Plato, Origen viewed Jesus as a cosmic soul, drawing all things into a coherent, living whole. To be alive is to participate in the world in a particular way. Origen also borrowed Aristotelian concepts of motion and nutrition to produce mortal souls and vegetable life. Spiritual life only aligns with these souls when they choose the proper path.

---

[11] In *Against Celsus* 7.39, Origen spoke of Genesis 3:7 after Adam and Eve ate from the tree of the knowledge of good and evil: "Then the eyes of both were opened." Their spiritual or intellectual eyes had always been open, but here their physical or aesthetic eyes were opened allowing bodily sight to cloud spiritual sight. Epiphanius of Salamis (c. 310–320–403) extended this metaphor claiming that God's gift of animal skins reflects the enfleshment of humanity. See Edwards (2014).

Christianity diversified an already wide field of life-concepts. Familiar ideas of vegetable souls and vegetable life persisted as physiological explanations for life and bodies in humans, animals, and plants. Those souls account for the circulation of blood and breath as well as nutrition and reproduction. The animal soul, or more simply the soul, remained the dominant way of explaining animal life: sensation and locomotion. The rational soul continued to distinguish humans with an ability to observe the Divine *Logos*.

In addition to these traditional categories, the New Testament added spiritual life. Humans can follow God and be born in Spirit through baptism. This spiritual life brings immortality. Humans who follow God will be granted eternal life in both body and soul through the Holy Spirit. Those who do not follow God receive eternal death in body and soul. Thus, the spiritual life is closer to *logos* or Plato's Wisdom than it is to any of the earlier life-concepts.

Meanwhile, the question of persistence through death shifted the language of souls away from material processes toward something immaterial yet substantial. Spiritual life took on the characteristics of Plato's star-souls, transcending the physical world, while calving off life-activities, which would become faculties or capacities for action. The soul remained the primary way of talking about the thing worthy of continued attention throughout biological change, the formal cause of a living thing.

References

Bynum, Caroline Walker. *The Resurrection of the Body in Western Christianity, 200–1336*. New York: Columbia University Press, 1995.

Cooper, John W. *Body, Soul, and Life Everlasting*. Grand Rapids, MI: Eerdmans, 1989.

Edwards, Mark J. "Origen." In *Stanford Encyclopedia of Philosophy*, Spring 2014 ed. Stanford University, 1997–. https://plato.stanford.edu/archives/spr2014/entries/origen/.

Green, Joel B. *Body, Soul, and Human Life*. Grand Rapids, MI: Baker Academic, 2008.

Gregory of Nyssa. *The Soul and the Resurrection*. Translated by Catharine P. Roth. Popular Patristics Series 12. Crestwood, NY: St. Vladamir's Seminary Press, 1993.

Heckel, Theo K. "Body and Soul in Saint Paul." In *Psyche and Soma: Physicians and Metaphysicians on the Mind-Body Problem from Antiquity to*

Enlightenment, edited by John P. Wright and Paul Potter, 117–131. Oxford: Clarendon, 2000.

Martin, Dale B. *The Corinthian Body*. New Haven: Yale University Press, 1995.

Martin, Raymond, and John Barresi. *The Rise and Fall of Soul and Self: An Intellectual History of Personal Identity*. New York: Columbia University Press, 2006.

Murphy, Nancey. *Bodies and Souls, or Spirited Bodies?* New York: Cambridge University Press, 2006.

New Oxford Annotated Bible: New Revised Standard Version with the Apocrypha (NRSV). Edited by Michael D. Coogan. New York: Oxford University Press, 2001.

Origen. *On First Principles*. Translated by G.W. Butterworth. New York: Harper and Row, 1966.

Origen. *Against Celsus*. Translated by F. Crombie. The Ante-Nicene Fathers 4. Grand Rapids, MI: Eerdmans, 1979.

Origen. *Homilies on Genesis and Exodus*. Translated by Ronald E. Heine. The Fathers of the Church 71. Washington, DC: Catholic University of America Press, 1982.

Tertullian. *De Anima*. Amsterdam: J. M. Muelenhoff, 1947.

Tertullian. *Treatise on the Soul*. Translated by Peter Holmes. The Ante-Nicene Fathers 1. Buffalo, NY: Christian Literature Publishing, 1885.

CHAPTER 9

Invisible Seeds: Life-Concepts in Augustine

Augustine of Hippo (354–430) marks the boundary between the Hellenistic world and medieval Christendom. Living in the final days of the Western Roman Empire, he had access to nearly a thousand years of natural philosophy. He corresponded with Christians and non-Christians from around the Mediterranean. A prolific writer, most Europeans came to see Plato and Aristotle through his lens, at least until the twelfth century. Augustine provided the hinge for two major shifts in thinking about life, each in response to divergent strands of Christianity. He defended the inherent goodness of both body and mind in response to Manichean dualism (i.e., matter is evil, mind is good). Humans experience salvation and resurrection in the flesh. He also defended God's agency in all things in response to Pelagianism (i.e., humans can will the good without Divine aid). God upholds all things, continuously. Augustine created a new Platonic synthesis of Christianity and Hellenistic philosophy, characterized by a dual creation, hylomorphic biology, and a subjective human soul with both intellect and will.

THE DUAL CREATION

Throughout his career, Augustine struggled with the two accounts of creation in Genesis. How might we reconcile them into one literal account of cosmic origins? The word "literal" had a slightly different meaning for Augustine. He wanted to know what really happened but recognized that face-value reading would be insufficient.[1] He attempted to tackle the problem in his early works *Against the Manichees* and *On the Literal Interpretation of Genesis*.[2] The latter proved too difficult and he left it unfinished. At the end of his life, he returned to the topic with a second volume by the same name. I use the Latin, *De Genesi*, when referring to the later work. The dual creation also appears in *The Confessions* (Chapters 12 and 13) and *The City of God* (Chapter 11). Though varying on some major points (e.g., human agency), the works agree on most topics related to vegetable life.

Philo had proposed dividing day 1 (eternal, invisible creation) from days 2 through 6 (temporal, visible creation). He treated Genesis 2 as an allegory. Augustine, unwilling to treat anything as purely allegorical, expanded the eternal creation (*De Genesi*, 5). Acting through the *Logos* (Christ for Augustine), God began in eternity with the creation of angels and cosmic archetypes (*rationes*). These archetypes are something like Platonic forms; they are atemporal and incorporeal. They explain the order of the universe. Because they exist eternally, they deal only with universal patterns and rules, not particular individuals. Augustine understood Genesis 1:5 as the the creation of 'day,' rather than the creation of a specific day. In a Neoplatonic sense, this is the mind of God. It is also the foundation for angelic and human intelligence. God makes the intelligible universe, in which intelligences are possible.

For Augustine, the next five days of Genesis 1 also refer to the *invisible creation*. God sets forth the details for specific kinds (e.g., trees, stars, humans). Time only makes sense in the context of a corporeal world; thus, the five days represent the hierarchy of creation, established in eternity. From unformed matter up through rocks and the various kinds of life to human intelligence, the cosmos follows this invisible order.[3] On the seventh day, God rested.

---

[1] The plain sense is, in some cases, clearly misleading, as when Genesis speaks of day and night before the creation of the Sun. See commentary in Augustine (1982, vol. 2, pp. 9–12).

[2] In full, *Two Books on Genesis Against the Manichees* and *On the Literal Interpretation of Genesis, an Unfinished Book*. Both are in Augustine (1991).

[3] Note that plants, stars, and angels all present difficulties for such a simplistic hierarchy of value.

The story continues with the *visible creation* in Genesis 2, wherein time and space begin. The sequence of events differs from Genesis 1, so it must represent a different aspect of creation. God manifested the visible cosmos according to the invisible plan. This visible creation was unbounded and continuous: God continues to create and sustain.

BIOLOGY

Looking at plants, the creation played out as follows. On the first day, God made the rules by which all living things operate: nutrition, reproduction, and development as abstract principles. Augustine believed that living things fall into clear categories with distinct rules for behavior and explanation (6.13–23). God made plants and animals from water and earth (3.2.3, 5.7.20, 7.12.18). Both perpetuate their type or kind through seed (6.6–16).[4] And hidden within all seeds is the power and potential for growth in living beings, including all stages of their lives. On the third day, God made plant *kinds* (8.3). This is neither the archetype nor a particular soul; it is the plan for souls and their activities within particular bodies. God works out the details of what it means to be a particular species. In Aristotelian terms, kinds are the potentiality and cause of a being that, in fullness, will be actual and dynamic (6.6.10).[5] They capture the way in which individual souls will act as efficient, final, and formal cause within living beings. In the Garden of Eden, God made individual plants to grow for the first time (Genesis 2:9). They sprung from the earth and started to perpetuate their kinds through visible seeds. Concrete, corporeal life and reproduction could then perpetuate kinds following the rules set forth on the first day.

The Hebrew word for kind, *miyn*, appears throughout the Hebrew scriptures, generally referring to a class of animals. It occurs ten times in Genesis 1, including three plant references (1:11–12). Augustine referred to the Septuagint (γένος) and uses the Latin term *genus*, still used in modern biology for groups of organisms (e.g., dogs as *Canis*, rice as *Oryza*). Building on Aristotle, he argued from heredity (offspring resemble parents) to the presence of form within the visible seed

---

[4] See also *The Nature and Origin of the Soul* I, 19–20.

[5] "invisibly, potentially, in their causes as things that will be in the future are made, yet not made in actuality now." Latin: "*Invisibiliter, potentialiter, causaliter, quomodo fiunt futura non facta.*"

(transmitting form from one generation to the next) to eternal form and kind existing in the mind of God (3.12.18–20).[6] Thus, the primordial reasons provide the invisible and eternal cause of plants, which God reveals over time.

These kinds are not the same as the forms in Aristotle. Though they are eternal in the mind of God, they need not be everlasting in the corporeal universe. Augustine cited the Biblical flood in Genesis 6–9. Afterwards, it may have been difficult or impossible for the descendants of ark animals to recolonize all the islands. God might have manifested plants and animals directly from the ground at different times (*City of God*, 16.7). Augustine also stated that some animals and plants arise normally through spontaneous generation. He spoke of their invisible seeds being interwoven through the bodies of plants and animals (*De Genesi*, 3.14.23).[7] When the latter decay, the seeds germinate and new creatures arise. Some invisible seeds may even be present in the soil, allowing new life to spring forth. Augustine's invisible seeds set the groundwork for high views of God's agency in plants during the Renaissance and Enlightenment.

THE HIERARCHY OF LIFE

Augustine explained souls along lines that should, by now, be familiar, in one place referring to all living beings, in another to animals, and most often to humans. Thus, he can say that anything alive is reckoned a soul (*Two Souls*, 6, Augustine 1990d). Though he is open to vegetable soul interpretations, he usually defines the soul as being a property of animals and speaks

---

[6]It is worth noting that Augustine followed Aristotle in saying that plants and (non-human) animals die and so reproduce themselves in pursuit of eternity. Originally, humans were capable of both reproduction and immortality, but because death was not certain, they had no strong drive toward sex and reproduction. This changed in the Fall (XI.32.42). See also V.23.44–46. The concept of primordial reasons is present in the Stoic literature. The Latin term *rationes seminales* is a direct translation of the Greek λόγοι σπερματικοὶ.

[7]God can make new individuals at any time, but all kinds are made in eternity (*De Genesi*, 5.20). Augustine did not comment on whether species may arise late in time or go extinct. Woods (2009) argues against species arising but seems to ignore his own admonition that the creation of species occurs in eternity and, therefore, there need be no temporal gap in which a species does not exist. Nor is God's manifestation of a species in time, of necessity, limited to the earliest days.

of plant natures (*natura*) with nutrition and reproduction. Something at work in plants differentiates them from non-living matter. This appears most clearly in his early work, *The Greatness of the Soul* (33.70).[8]

> [The] soul by its presence gives life to this earth- and death-bound body. It makes of it a unified organism and maintains it as such, keeping it from disintegrating and wasting away. It provides for a proper, balanced distribution of nourishment to the body's members. It preserves the body's harmony and proportion, not only in beauty, but also in growth and reproduction. Obviously, however, these are faculties which man has in common with the plant world; for we say of plants too, that they live, [and] we see and acknowledge that each of them is preserved to its own generic being, is nourished, grows, and reproduces itself.

Animals are identified by sensation and spontaneous motion, while humans are mortal rational animals. The hierarchy appears succinctly in *On the Free Choice of Will* (1.8.18) and, in more detail in *The City of God* (5.11):

> To good and evil men alike He gave being, in common with the stones; and He gave life capable of reproducing itself [*vita seminalis*], in common with the trees; and sentient life [*vita sensualis*], in common with the beasts; and intellectual life [*vita intellectualis*], in common with the angels alone. … to the irrational soul He has given memory, sensation, and appetite; and to the rational soul He has in addition given mind, intelligence and will.

Augustine rarely spoke of plants directly, being more inclined to speak of things bearing seed, lower life, or specifically trees. All life is characterized by internal motion. In other words, organisms motivate themselves. In the case of simple life, internal forces drive nutrition, growth, circulation and reproduction (*De Genesi*, 7.16.22). And this parallels the same functions driven by the soul in animals, specifically blood circulation and the growth of hair and nails. Plants, however, lack senses, differentiating them from animals (3.13.21). For this reason, and because they lack locomotion, they belong more to the ground than to the class of living things (2.27). This explains their creation on the third day and not the sixth.

[8]Taken from the Colleran 1950 translation, quoted in Goetz and Taliaferro (2011, pp. 32–33).

Augustine said a great deal more about animal life. The Latin word *animale* literally means "ensouled" and refers to living creatures which sense or breathe. When he wished to more clearly identify non-human animals, he referred to brutes, irrational animals, or livestock. Rejecting the Manichean notion that humans possess both a good soul (of God) and a bad soul (of Darkness), Augustine went out of his way to affirm the goodness of animal nature, even the goodness of all souls. In Chapter 4 of *Two Souls* he waxed poetic on the dignity of fly souls. That which organizes and motivates an organism, even one so minute, must be of great value, even greater than that of light, the most rarified and exalted part of inanimate creation. It would be hard to have a clearer endorsement of non-human souls.

Augustine's souls exist in and of themselves (by God's will) but are immaterial (*De Genesi*, 7; *Nature and Origin of the Soul*, 4.18). We can know this because, in sensation and memory, they hold incorporeal images. Even irrational animals know how to find their way home (*De Genesi*, 7.20.26). We know the soul lacks extension because it co-exists with the body. (Bodies change in space and time; souls in time alone; God not at all.) Despite dealing with questions of sensation and locomotion, the explanation differs greatly from Aristotle's concept; Aristotle's souls were processes in action and in fulfillment, but Augustine's were substantial entities that grant power. Thus, we have souls metaphysically Platonic but functionally Aristotelian. We have souls and their *faculties*.

Animal souls have several faculties, the most prominent of which are sensation and spontaneous motion. Fire and air mediate between the immaterial soul and the material body, which springs from earth and water. The more rarified elements more closely resemble spiritual substance and can carry signals to and from it (3.4.6–7, 7.15.21). Sensation is familiar and resembles Aristotle's *aesthesis*. Spontaneous motion is more problematic. Augustine perceived a hierarchy of agency. Animals, with "internal motivation," have more freedom than plants. Humans, with "will," have more freedom than other animals. Augustine also associated three internal faculties with the animal soul. The inner sense integrates sensory input; the memory recalls previous sensations; and the appetite judges favorable and unfavorable sensations based on the proper ends of vegetable and animal souls.

THE INNER PERSON

Gareth Matthews (2000, p. 135) speaks of a new dualism that arose within Augustine's works based on self-reflective arguments. This subjective self had a profound influence on later notions of the mind and soul. The inner self is not only immaterial but ontologically and epistemologically distinct from the sensible world.

A millennium before Descartes, Augustine argued from cognition to existence. He added a twist that neatly frames the epistemic problem (*On the Free Choice of the Will*, 2.3; *On the Trinity*, 10.14; *City of God*, 11.26). I think I exist and, if I err in thinking that, who has made the mistake? Rather than logically moving from premise to conclusion, this *Dubito*, or doubt-based argument, highlights the seemingly necessary presumption that one is a distinct individual agent. Who am I? I am the one asking the question. I am the one who may succeed or fail in finding a satisfactory answer. Augustine went on to say that this realization leaves us no choice but to recognize our own existence as a subject.[9] In this, he was solidly Aristotelian. As rational beings, our existence *is* our thinking, not our presence as an entity capable of thought.

Augustine inferred, by analogy, the existence of other subjects. If I exist as a thinking self, other beings, similar in appearance, must as well. He used several words to describe this aspect of humanity. With Lucretius and Tertullian, he invoked a masculine *animus* (e.g., *The Nature and Origin of the Soul* 1.19 and *Against the Manichees*, 1.17). He also spoke of the *mens* as both subject and object of perception (e.g., *De Genesi*, 12.10.21). The easiest translation for either is "mind," which functions in both intellect (*intellectus*) and reason (*ratio*). By intellect, the mind perceives the invisible order of creation, just as the eyes perceive the visible form of matter. Participation in the invisible creation gives it uncorrupted knowledge of a thing in itself. This has variously been referred to as revelation, intuition, and a priori knowledge, though none of those completely express the original sentiment. Alternatively, reason captures the fallible process of moving from perceptions (intellectual and sensual) to conclusions. Augustine saw intellect and reason essentially integrated as functions of the same mind.

---

[9] Matthews (2000) points out that this conclusion is not necessary. Heidegger and Wittgenstein both provide famous critiques of the Cartesian version, while Buddhism and other Asian philosophies do not even subscribe to the basic categories.

The intellect grants human souls (and angels) the ability to perceive the invisible creation directly, unmediated by the physical body. Thus, Augustine could speak of three kinds of vision that parallel the vegetable, animal, and rational souls (*De Genesi*, 12.11.22). Corporeal vision, mediated by the physical eye, reveals the letters on the page. Spiritual vision (here meaning spirited or animate rather than Divine) shows the letters to the mind's eye. They can remain there, even after the physical image has gone.[10] Non-human animals possess both corporeal and spiritual vision. Intellectual vision reveals words and the meaning behind them. The soul can, of course, be mistaken (12.25.52). The soul is misled by corporeal vision when the eye malfunctions, as when a fever makes the letters swim or when people are colorblind. The soul is misled by spiritual vision when we misinterpret what we see, as when the brain automatically edits out a repeated word or typo. The soul cannot be misled by the intellect.

Many things may be known with only an appeal to the eyes and the reason. The frog is green. The stars move relative to the ground. Some things, however, require the use of intellect. To say that all frogs are green, we will need a category or "kind" that has common properties. To say whether it is the stars or the earth that move (in an absolute sense) requires a reference frame. Augustine's account of life and creation only operates within this frame. The rules and kinds of biology can only be perceived by a mind with intellect (6.16.27). And the mind itself can only be perceived by another mind (12.10.21). The vegetable and animal soul, in Augustinian thinking, are not open to physical observation. They can only be known by the intellect.[11] As the visible and invisible creation drifted apart during the Middle Ages, as the physiological and psychological life-concepts divided, it would become harder to make sense of life-concepts along the traditional lines. Both Greek and Christian traditions depend critically on a psychic faculty for perceiving immaterial reality.

---

[10] This idea draws on the *phantasia* or imagination in Aristotle, likely through the mediation of Origen.

[11] Strictly speaking, the motion caused by vegetable and animal souls can be seen with the senses. By this motion we know the principle of motion behind it must exist as well (*On the Trinity*, 8.6.9). The intellect is required to perceive the ordered regularity of the kinds and invisible seeds.

In *On the Trinity*, Augustine made a somewhat strained argument for the immateriality of the mind based on our self-knowledge (Matthews 2000, p. 141; see also Goetz and Taliaferro 2011, pp. 35–44). Of course, this only gets us to immateriality for the human soul and not for animal and vegetable souls. Irrational souls pass from generation to generation through the seeds, a traducian position. If they are immaterial, they are necessarily tied to material (and physical) expression. Augustine remained ambivalent about the origin of human souls. He favored the "special creation" of human souls (by God at the time of ensoulment), but worried about original sin. Why would God create new souls with the old stain? Thus, he never gave up the possibility of traducianism for humans as well (Letter 166; *On the Soul and Its Origin* XVII.14).

The Rational Animal

For Augustine, the criteria of mortality, animality, and rationality were not only necessary but sufficient to humanity.

> But whoever is anywhere born a man, that is, a rational, mortal animal, no matter what unusual appearance he presents in color, movement, sound, nor how peculiar he is in some power, part, or quality of his nature, no Christian can doubt that he springs from that one protoplast. We can distinguish the common human nature from that which is peculiar, and therefore wonderful. (*City of God*, 16.8)[12]

Mortal and animal require no further discussion. It is unclear whether Augustine counted angels and demons among the animals. *De Genesi* suggests that angels, at least, are invisible, incorporeal, eternal creatures of reason. Near the beginning, however, he flirts with the idea that angels and demons are animals of air and fire (*De Genesi*, 3.9–10.13–15; see also notes 60, pp. 238 and 37, p. 244). If so, the highest angels are composed of a rarified fire, like the aether of Aristotle. Demons, fallen from among the lower angels, were transmuted into damp air, making them subject to punishment by fire. Augustine vacillated on the question

---

[12] In the same section, he specified—whether they exist potentially or actually—cyclopes, hermaphrodites, antipodes, skiopodes, pigmies, the mouthless, the headless, the short-lived, and those with dog-heads; so long as they are rational and mortal animals, they are children of Adam.

of stars and planets, whether they be rational or animal (*Enchiridion*, 58; *De Genesi*, 2.18, see note 63, p. 239). In any case, angels, demons, and stars are immortal.

THE WILL

Having established an intellect and an inner self, Augustine could distinguish between the agency of humans and brutes. Humans have will (voluntas). By will, humans can choose to follow their intellect—or not. Spiritual life in Augustine requires both; all humans have both; but, only those who use them together can find a second birth and true resurrection.

At first, humans used their wills in accord with intellect and Divine will. The flesh was subject to the intellect, mediated by body and soul (following Plato). Human intellect, perceiving the mind of God without impediment, ruled. By an act of will, the first humans chose not to follow the intellect, breaking the proper order within themselves and within creation.[13] This was the first sin and brought about the Fall. Sin requires a will, not that humans will to sin, but that they will an act which is objectively out of alignment with the mind of God ("One Book Against the Manicheans" *Revisions*). The subjectivity of the self, the inner person, allowed Augustine to distinguish between the soul as perfect philosophical guide and soul as charioteer, resolving some of the ambiguity in the Platonic soul. At the same time, it created an epistemological gap between self and other. Not only is sense data unreliable because of the changeability of matter (the Platonic argument), it is "intractably private" (Mendelson 2012). In separating the self from the cosmic organism of Neoplatonism, Augustine has also created an intractable gap between the perceptions of different people.

Drawing on Philo and the Stoics, Augustine described an internal conflict. The soul and the will, placed in the middle, are drawn either to

---

[13] *De Genesi* XI.32.42. This loss resulted in corrupted vision and a desire to reproduce. Augustine's early works suggest that it also brings about mortality (*Against the Manichees* II.21), though his position becomes more sophisticated in later writings. In *De Genesi*, humans are potentially immortal in creation, but not actually immortal until the second resurrection. Note that death has come from sin, but mortality is simply the juvenile state of humans; maturity brings immortality (*De Genesi* VI.22–23.33–34). See *City of God* XIII.3 for an intermediate view.

the spirit or to the flesh (*Confessions*, 8.5.10). There is only one human (physical and spiritual) and yet, the human can favor an internal or external aspect of its existence. Throughout, Augustine sought to explain the divide within humans. Why do we feel such conflict? He recognized in himself a tendency to choose badly, even when his intellect showed him the right thing to do (*De Genesi*, 5.14.31). And so, subjective life diverged from rational and spiritual life.

Two Deaths, Two Resurrections

Augustine reinterpreted questions of death, soul, and resurrection, away from physiology and toward redemption from sin. In the *City of God* (20.6), he spoke of two deaths and two resurrections. In Adam's Fall, all souls died; in baptism, some souls are resurrected. At the last judgement, all souls will be resurrected in bodies; unbaptized, some souls will die again along with their bodies. Neither death refers to physiological death as commonly understood.

Later in life, Augustine shared a different division of life-concepts in *De Genesi* (12). Wicked souls enter the *Inferno* (literally underworld), a substantial but incorporeal place of eternal punishment, corresponding to the Greek Tartarus and the lake of fire in Revelation. They suffer from the spiritual likeness of flesh, as one might dream of physical injury (12.32.61).[14] Blessed souls find reward in a three-fold paradise, with corporeal, spiritual, and intellectual levels, each more rarified than the last. In the intellectual heaven, the just can perceive all objects unimpeded by the flesh.

Both schemes are difficult to align with ideas of physical resurrection.

Augustine solidified the dual creation first proposed by Philo. In it, he distinguished between the eternal realm of divine intellect and the temporal realm of physical creation. This divide is not yet the dualism of Descartes, but we can now see it on the horizon. Physiological and subjective accounts must be different because subjective explanations involve an incorporeal substantial soul.

Plant life follows rules set down on the first day. The principles of internal motion and motivation, including nutrition, growth, and

---

[14]Compare *City of God* 21.10, where humans suffer the fire of punishment in their earthy bodies, while demons suffer in airy or spiritual bodies.

reproduction are present in all living things. They are organisms because they are organized according to God's plan in eternity. They exist in kinds, normally passed on from generation to generation by seeds, though some few may be made in special acts of creation or through invisible seeds waiting in matter until the proper time.

Vegetable natures explain nutrition, reproduction, and the organization of living beings. Though Augustine does not reflect on them specifically, the mechanics of soul/body interaction suggest that he saw them as immaterial, but substantial, just like the human soul. They need to explain the same sorts of things within plant bodies that human souls explain in human bodies, including circulation and development. Similarly, animal souls continue to explain sensation and locomotion. Augustine seems to vacillate between thinking of uninspired humans as brutes and thinking of them as immortal souls destined for eternal death.

The now familiar ambivalence about a unity of souls has been partially resolved by making the soul an indivisible substance responsible for animal life. Just as vegetable life has been dropped below the soul threshold into plant natures, kinds, and invisible seeds, so rational and spiritual life have been raised above the human soul. Intellect (as true perception of the mind of God) and spirit (as participation in the Holy Spirit) require something greater than human mind.

This medieval synthesis on Platonic lines emphasized God's action uphoding biological order through transcendent souls and plant natures. An alternative synthesis along more Aristotelian lines emphasized the operation of souls in building up matter.

REFERENCES

Augustine. *The City of God*. Translated by Marcus Dods. New York: Random House, 1950.

Augustine. *The Greatness of the Soul, The Teacher*. Translated by Joseph M. Colleran. Ancient Christian Writers Series. Mahwah, NJ: Paulist Press, 1978.

Augustine. *On the Literal Meaning of Genesis*, 2 vols. Translated by S.J. John Hammond Taylor. Ancient Christian Writers Series 41–42. Mahwah, NJ: Paulist Press, 1982.

Augustine. "Letter 166." In *Letters 156–210*, translated by S.J. Roland Teske, 77–93. The Works of Saint Augustine: A Translation for the 21st Century. New York: New City Press, 1990a.

Augustine. "The Nature and Origin of the Soul." In *Answer to the Pelagians*, translated by S.J. Roland Teske, 465–561. The Works of Saint Augustine: A Translation for the 21st Century. New York: New City Press, 1990b.

Augustine. "One Book Against the Manicheans on the Two Souls." *Revisions*. In *Works of Saint Augustine: A Translation for the 21st Century*, translated by Boniface Ramsey, 69–78. New York: New City Press, 1990c.

Augustine. "The Two Souls." In *Manichean Debate*, translated by S.J. Roland Teske, 106–134. The Works of Saint Augustine: A Translation for the 21st Century. New York: New City Press, 1990d.

Augustine. *St. Augustine on Genesis: Two Books on Genesis Against the Manichees and On the Literal Interpretation of Genesis: An Unfinished Book*. Translated by S.J. Roland J. Teske. The Fathers of the Church 84. Washington, DC: Catholic University of America Press, 1991.

Augustine. *On the Free Choice of the Will, On Grace and Free Choice, and Other Writings*. Translated by Peter King. New York: Cambridge University Press, 2010.

Augustine. *Sancti Aurelii Augustini Opera Omnia: Patrologiae Latinae Elenchus* online. Accessed July 2017, http://www.augustinus.it/latino/index.htm.

Goetz, Stewart, and Charles Taliaferro. *A Brief History of the Soul*. Malden, MA: Wiley-Blackwell, 2011.

Matthews, Gareth. "Internalist Reasoning in Augustine for Mind-Body Dualism." In *Psyche and Soma: Physicians and Metaphysicians on the Mind-Body Problem from Antiquity to Enlightenment*, edited by John P. Wright and Paul Potter, 133–145. Oxford: Clarendon, 2000.

Mendelson, Michael. "Saint Augustine." In *Stanford Encyclopedia of Philosophy*, Winter 2012 ed. Stanford University, 1997–. http://plato.stanford.edu/archives/win2012/entries/augustine/.

Woods, Henry, S.J. *Augustine and Evolution: A Study in the Saint's "De Genesi ad Litteram" and "De Trinitate."* Eugene, OR: Wipf and Stock, 2009 (1924).

CHAPTER 10

Aristotle Returns: A Second Medieval Synthesis

The Platonic and Augustinian view of life as participation in the cosmic *logos* dominated biological thinking through the early Middle Ages and into the twelfth century. It emphasized immaterial agency and top-down causation. An alternate perspective arose, running through some of the most influential thinkers of Islam, Judaism, and Christianity. An Aristotelian synthesis worked with substantial souls and material dynamics. In place of three-fold souls, it posited single souls with three-fold *faculties* or abilities. All of Aristotle's causes took on new definitions, radically changing theories of life and agency. The two syntheses were nonetheless deeply entwined. The Platonic synthesis drew heavily on Aristotelian and Stoic terminology. Similarly, the rediscovery of Aristotle came through the lens of Neoplatonic commentary. The new perspective highlighted souls as individual and subsistent. It clarified the role of vegetable souls and faculties but made animal faculties more problematic.

Aristotle Reinterpreted

Aristotle's works were available to scholars in the Golden Age of Islam, particularly in the cultural centers of Baghdad and Cordoba. They were, however, transmitted through commentaries that blend copying, summarizing, and improving the original. This resulted in a shifted "Aristotelianism."

© The Author(s) 2018
L. J. Mix, *Life Concepts from Aristotle to Darwin*,
https://doi.org/10.1007/978-3-319-96047-0_10

Commentaries by Alexander of Aphrodesias (ca. 300) and Themistius (317–ca. 390) emphasized the immaterial character of human souls. Both focused on the intellect, but they did so in different ways. Alexander spoke of a singular immortal and immaterial Agent Intellect, which informs the passive intellect of all humans (Martin and Barresi 2006, pp. 24, 78). Individual passive intellects are mortal; only the Agent Intellect survives death as an immaterial entity. Alternatively, Themistius recast the human intellect as an individual essential self (Martin and Barresi 2006, pp. 78–79; Sorabji 2010). Over the centuries, Aristotelian thinkers vacillated between these two perspectives on the rational soul.

Pseudo-Simplicius (sixth century) changed the character of final causes, shifting the emphasis from organization (i.e., having coordinated parts) to instrumentality (i.e., being used for something) (Blumenthal 1996, p. 94). Final cause and "in fulfillment" changed from intrinsic motivations to extrinsic purposes. Specifically, vegetable and animal souls became tools, either for God or for rational human souls. Pseudo-Simplicius also reframed the work of the vegetable soul from a species aimed at eternity to matter seeking the divine (Simplicius 1995, p. 148). The consistent Neoplatonic hierarchy can obscure the dramatic change from being flowing downward to matter striving upward.

Ammonius (435–517) and his student John Philoponus (490–570) also reinterpreted final causes, describing them in terms of intention and need (Leunisson 2010).[1] Philoponus explicitly divided body and soul. He said that the human soul, with regard to the body, is the dynamic form of a corporeal entity but, with regard to itself, is not an activity but an independent substance (Hasse 2008). Thus, the human soul became an incorporeal cause of corporeal activity. I will refer to such souls as *subsistent* because they support themselves without either material cause (parts of which they are made) or matter (stuff which they shape). Subsistent souls inform matter but are more than the form of matter. They provide the potential or power for vital activities and exist prior to them.

While the subsistent soul arose in Neoplatonism, it must be distinguished from the participatory soul of the Platonic synthesis. From Philo to Augustine, the soul was integrally wrapped up in the life of

---

[1]Ammonius wished to reconcile Aristotle's prime mover with Plato's demiurge in the *Timaeus*. Philoponus wished to reconcile the prime mover with the Christian God.

the cosmos. In the Aristotelian synthesis, souls became the units from which explanations are built. The difference rests in where we start our explanations, from universals or from particulars. It must be noted that Philoponus did not believe in subsistent vegetable souls (Philoponus 2005, p. 32). The vegetable faculties can only be expressed in a gross physical body; therefore, they die when the body dies. He notes, though, that hair and nails continue to grow after death and that the body begins to decay. Thus the vegetable soul, or the vegetable aspect of the soul, persists in a minor way after bodily death.

Philoponus, a Christian, associated spiritual life with immateriality (pp. 33–36). The rational soul can be purified of its corporeal attachments (including animal and vegetable faculties) and be saved. Rational souls that do not achieve this goal flee to Hades, taking an animal soul with them. In Hades, a pneumatic body exists as a rarified collection of elements. Thus physical, it can experience pain and, therefore, punishment. It can also return to the mortal world as a ghost, bereft of common matter but still corporeal and capable of interaction.

THE FALSAFAH

The first self-styled philosopher of the Islamic world was Ya'qûb ibn 'Isḥâq Al-Kindî (ca. 800–870). Educated in Baghdad and sponsored by 'Abbasid caliphs, he translated numerous works from Greek into Arabic, rendering *psyche* in its various uses with the Arabic word *nafs*. He merged Aristotelian "first philosophy" with theology, seeing God as the first cause of all things (Adamson 2015). In his *Discourse on the Soul*, al-Kindî spoke of the three-part soul but distinguished between the vegetable and animal faculties, which are activities of the body, and rational faculties, which are "the intellectual form of the living thing." He likewise divided sensual and intellectual perception. Thus, from its earliest roots, the Aristotelian synthesis viewed the animal/rational divide as the most important distinction in the study of life. Vegetable and animal faculties, following Aristotle, are "perfections" of living bodies. They are embodied processes driven by physical organs. The intellect is an incorporeal faculty possessed by the subsistent human soul.

Followers of al-Kindî worked to reconcile Hellenistic life-concepts with the doctrines of Islam while developing new medical and biological sciences. The medical pioneer al-Râzî (854–925) speaks of three souls (Ivry 2012). The vegetable and animal souls are physical, located in the liver and heart respectively, while the rational or divine soul acts through the brain to order them.

Abû Naṣr al-Fârâbî (870–950), a philosopher active in Baghdad, had a much larger impact on the development of life-concepts. He elaborated Aristotle's hierarchy of biological activities. With al-Kindî and al-Râzî, he noticed a significant difference between the proto-agency of animals and the intellect-informed agency of humans.[2] In his *Enumeration of the Sciences*, al-Fârâbî listed the biological activities as nutrition, sensation, appetite, imagination, and reason (López-Farjeat 2016). His list warrants a closer look. The vegetable faculty is still labeled "nutrition" and includes growth and reproduction as well as the processing of nutrients. The animal faculty has been expanded. Sensation remains the same, but al-Fârâbî spoke of appetite as the aspect of agency that allows animals to be attracted or repelled by what they sense, often based on its value for nutrition and reproduction. Thus, locomotion becomes willed motion. The appetitive faculty should not be confused with the appetitive soul of Plato, because here it requires sensation and a greater ability to move toward or away from what is sensed.[3] Thus the problem of locomotion fell within appetite for al-Fârâbî. Likewise, the imagination—sometimes reserved to the rational soul by later authors—allows animals to remember and anticipate attractive and repulsive sensations and navigate the world.

Both al-Kindî and al-Fârâbî followed Alexander of Aphrodesias in denying individual immortality (Martin and Barresi 2006, p. 82). God has an immortal intellect; humans are immortal only in their participation with the Divine. This denial of individual spiritual life after death met heavy resistance in Islam, as in Judaism and Christianity.

Ibn Sînâ

Abû-'Alî al-Ḥusayn ibn Sînâ or Avicenna (ca. 970–1037) was the most influential proponent of *falsafah*, Islamic philosophy. The Samanid Emir in Bukhara sponsored him early in his career, but political troubles made

---

[2] Al-Fârâbî used the terms *irâdah* for animals and *ikhtiyâr* for humans (Ivry 2012).

[3] Al-Fârâbî drew on Aristotle's division of the soul in the *Nichomachean Ethics* (I.13), where he identified the animal soul with appetite. Aristotle also mentions the appetite as such in the final chapters of *On the Soul*, 3, though it is less clear here that it fits neatly into the animal soul.

him itinerant after 999. Muslim thinkers still recognize Ibn Sînâ as the height of philosophical thinking in Islam, though philosophy became much less popular in the following centuries.

Ibn Sînâ started with Aristotle's dictum that the soul is the first actuality and the second potentiality of life (Ibn Sînâ 1906, sec. 2). It is both organized matter and the principle by which organized living beings act. Ibn Sînâ interpreted this to mean that matter is insufficient for life. Living things must involve an additional substance. The soul, which was an identity of efficient, formal, and final causes in Aristotle, was reframed as a component. This required an adjustment of all four Aristotelian causes. The material cause, once any appeal to composition, became a kind of stuff—namely the elements. This new understanding of souls and material causes undergirds the modern concept that body and soul are, in some way, equivalent but ontologically distinct pieces which come together in life. One is a physical substance, in modern as well as falsafah thinking; the other is a non-physical substance. The efficient cause remained the same, though it worked differently, as detailed below. The formal cause included these new metaphysical substances, something akin to Platonic forms, but without the grand narrative of becoming and being. These forms subsist as proto-agents with faculties, the potentiality for actual living activities. The final cause also shifted. When the soul changed from activity to proto-agent, the final cause changed from a dynamic perfection (in-action and in-fulfillment) to a static goal. All of these interpretations dramatically altered both life-concepts and souls.

Ibn Sînâ classified explanations about life in a way more familiar to modern readers, if still somewhat strange. Following Aristotle, he viewed natures (Greek *physis*, Arabic *ṭabîʿa*) as an internal cause of motion.[4] And yet, he thought of natures as agential, efficient causes. They are that which, in a thing, directs it to move. With this new focus, he divided all activities into four categories with two criteria: uniformity and volition (or will).[5] He attributed non-volitional uniform motion to nature. This included the movement of elements, what we now think of as physics.

---

[4]McGinnis (2016) citing *The Cure* I.5. McGinnis details the significant differences between Aristotle's use of 'nature' and Ibn Sînâ's.

[5]Volition here is the same *irâdah* used by al-Fârâbî to describe the lesser agency of non-human animals. Elsewhere, Ibn Sînâ attaches it more closely to the type of agency unique to humans.

He attributed non-volitional non-uniform motion to vegetable souls. Plants grow irregularly but without will. Volitional non-uniform motion belongs to animal souls. Animals seek out some things and avoid others. Finally, volitional uniform motion belongs to celestial souls, possessed of intellect and moving in regular circles through the heavens. Humans are animals. They possess an intellect but lack the regular motion of the stars.

Greater detail on vegetable activities can be found in *The Compendium on the Soul*, *The Cure*, and *The Canon of Medicine*. The most fundamental activity of vegetable souls is nutrition, "that whereby the aliments are transformed into the likeness of the thing nourished, thereby replacing the loss incidental to the process of life" (Ibn Sînâ 1973, p. 113: *Canon* 1.6.2). Nutrition is a necessary precondition for growth and reproduction. The former is always present; the latter happens sporadically. This makes nutrition the basic activity of vegetable souls. And yet, it is not etiologically primary because nutrition is ultimately for the sake of propagation, recalling Aristotle's idea that mortals seek eternity by reproduction (*Compendium*, sec. 4). Plants have four subsidiary activities powered by the vegetable soul: attraction, holding, digestion, and excretion (ibid.; see also *Canon* 1.5.9, 1.6.3). In other words, each plant can move and sort elements while processing food. Ibn Sînâ makes clear that all the vegetable faculties work through the natures of elements and the primary qualities of hot, cold, wet, and dry (*Canon* 1.6.3).[6] The action of elements serves the vegetable faculties. In humans, the vegetable faculties drive from the liver (nutrition) and sex organs (reproduction).

In *The Cure*, Ibn Sînâ devoted the seventh book on natural science to vegetables. He defined life as a capacity for voluntary movement, denying that plants meet the criterion.[7] He distinguished vegetables from minerals with a unique faculty, present in all earthly living things: nutrition or the conversion of that-which-is-not-self to self. This expressly involves capturing or recruiting elements to serve the needs of the organism in propagation. Just as medieval lords war for territory (and the

---

[6]Philoponus discusses precisely this in his commentary on Aristotle's *On the Soul*; see Blumenthal (1996), p. 119.

[7]Tawara (2014). This echoes the skepticism about plant life in *On Plants*, attributed to Aristotle, but likely written by Nicolaus of Damascus. A translation of the latter work can be found in Barnes 1984.

farmers on it), so souls compete for heat and elements to sustain their actions. While he was confident that vegetables have souls and require a unique means of explanation, he hesitated about whether to call them truly alive. He associated life with physical breath, Divine breath, and volition.

In any case, humans and other animals possess the vegetable faculties necessary for their survival. Emphasizing the importance of unitary souls for explanation, Ibn Sînâ argued that humans have a single soul, which possesses vegetable, animal, and rational faculties. Each level serves the higher level. Thus, nature serves the vegetable, the vegetable serves the animal, and the animal serves the rational. In a similar way, discrete plants serve animals and animals serve humans. The animal faculties, as we have seen in earlier authors, are driven from the heart and associated with blood and breath (*Compendium*, secs. 5–7). Chief among them are sensation and locomotion. The animal faculties extend to the outward senses (touch, taste, smell, hearing, vision) and to the inward senses, which he attributed to specific regions of the brain (Ivry 2012). A *common sense* coordinates the outward senses. Ibn Sînâ added memory, imagination, and estimation as ways of digesting sensory data. He also attributed cogitation and judgment to the animal faculties, but only within rational animals. Vegetable and animal souls die with their bodies, as do the vegetable and animal faculties of human souls (*Compendium*, sec. 10). Without matter to operate through, they cannot be active and, thus, are no longer meaningfully called faculties.

Finally, Ibn Sînâ referred to the mind. Through the rational faculty and intuition, the mind can participate in the divine Agent Intellect, the "Giver of Forms," and know truth (Martin and Barresi 2006, pp. 83–84). The mind receives images through the animal faculties. It can then reason upward to universal truths. Thus, reasoning has been smeared over a range that includes God (spiritual life), the human soul (rational life), and the sensitive faculty (animal life). The whole spectrum will be key to understanding scholastic and Enlightenment epistemological debates.

Ibn Sînâ provided his own argument for personal existence, known as the Floating Man. In *The Cure*, he provided a thought experiment, wherein a man is floating in dark space without touching or seeing anything. Such a man still has, he says, a sense of self, proving the existence of a mind that transcends animal or sensitive life. With previous Islamic philosophers, he identifies this as the passive intellect. He grants it personal immortality, but only as an immaterial collection of knowledge participating in the mind of God, not as a subsistent agent.

AL-GHAZÂLÎ

Abû Ḥâmid Muḥammad Al-Ghazâlî (ca. 1058–1111) holds a special place in the history of Islamic thought, living at the inflection point where philosophy and scholastic-style theology give way to more deontological reasoning. He found Ibn Sînâ's philosophy largely consistent with Islam but insists that Ibn Sînâ and Aristotle were wrong on three points of doctrine (*The Incoherence of the Philosophers*, Al-Ghazali 2000). First, he defended the division of Creator and creation; matter cannot be co-eternal with God. Second, he claimed that the mind of God can attend to individuals, not just universal principles. Third, he asserted physical resurrection. All three of these destabilized Ibn Sînâ's view of the resurrection.

Al-Ghazâlî also argued for occasionalism, the doctrine that every instance of cause and effect is an occasion for Divine action. Nothing occurs without God immediately willing it. *Falsafah*, he said gives too much weight to efficient causes. One might argue that both Ibn Sînâ and Al-Ghazâlî desired regularity within a universe constantly upheld by God. *Falsafah* emphasized that the world is comprehensible to insulate the perfect, unchanging God from changeable nature (recall Parmenides and Heraclitus). Al-Ghazâlî emphasized God's ability to act through miracles against the predictions of natural philosophy.

Ibn Rushd

'Abū l-Walîd Muḥammad ibn Rushd or Averroes (1126–1198) provided a rearguard for the Islamic philosophers. As a Maliki legal scholar, his opinion was influential in Islam, but his philosophical positions gained little traction. His rebuttal of Al-Ghazâlî, however, provided ammunition for Thomas Aquinas and the ascendance of Aristotle among Christians. With regard to vegetable souls and reasoning, his positions matched closely with those of Ibn Sînâ. He pushed back against physical resurrection, going even farther than Ibn Sînâ. Individuality requires a fusion of sensitive and rational faculties and ends with bodily death (Martin and Barresi, p. 84). Only the Agent Intellect exists eternally.

MAIMONIDES

In medieval Judaism, Rabbi Moses ben Maimon (also known as Rambam and Maimonides, 1138–1204) picked up the torch. Living in Muslim Cordoba and later in Morocco and Egypt, Maimonides had access to Greek works as well as Ibn Sînâ and other contemporary scholarship.

His early writing supported an Aristotelian vegetable soul.[8] The human soul is unitary with five faculties: Nutritive, perceptive, imaginative, appetitive, and rational (identical to al-Fârâbî's list). The vegetative aspect of the soul—present also in vegetables—can be expanded to seven faculties, following Ibn Sînâ: attraction, retention, digestion, excretion, growth, reproduction, and differentiation. Perception, imagination, and appetite fell to the animal soul (as in Ibn Sînâ) along with the elaboration of internal and external faculties. The rational soul again provided access to eternal truth, unavailable through physical perception.

Maimonides went out of his way to note that the nutritive faculty of plants is not the same as the nutritive faculty of humans. I believe this is most easily explained as a defense of psychic unity; it prevents opponents from claiming that humans have three souls, one identical to that which motivates plants in addition to the animal and rational souls. Michael Marder (2014, p. 104) rightly notes that this has the effect of diminishing plants in terms of intrinsic worth and ethical consideration.[9]

Maimonides' most famous work, *The Guide for the Perplexed* (Maimonides 1904), provided a second look at plants and souls, this time more in line with Hebrew Scriptures. Here, he defined life (*chay*) as sentience and wisdom (1.42). The soul (*nephesh*), likewise, he associated with blood, sentience, will, reason, and persistence after death (1.41). Life, will, and thought all proceed from the heart (*leb*, 1.34). He found none of these in plants.

Maimonides also associated life with organismality, the working of parts for the good of the whole (1.62–63). With Neoplatonic participation weakened, biological organization once again required an explanation. Maimonides linked life with parts working for the good of a whole. This did not bode well for plants. The heart is the source of life, and

---

[8]For vegetable souls, see the preface to *Mishnah Avot* in *Commentary on the Mishnah* and *Eight Chapters*, Chapter 1.

[9]I am not convinced by Marder's further claim that Maimonides was motivated by his principle against the mixing of kinds. The avowal of psychic unity seems sufficient.

parts of living things should die when sundered from it. Animal organization can be seen in limbs that die when cut off from the core. Plant cuttings survive. This became an important point for Enlightenment and Romantic reasoning about plants; they are indiscrete. Despite the emphasis on individuals, Maimonides still considered the cosmos to be a living animal. Its heart and soul are God (ibid.). Thus, all things die when they are severed from the Divine. The planets are also animals, moved by souls and intellects, as in Ibn Sînâ (2.4–5).

Maimonides made a distinction regarding will and morality. He stated that the nutritive and imaginative faculties are value free (*Eight Chapters*, Chapter 2, Maimonides 1912). They aim at the physical needs of organisms without any sense of right or wrong. Similarly, the intellect, when limited to the faculty for perceiving the eternal, must be value-free. It only perceives truth. The other animal and rational faculties, however, in perceiving what is good (by means of the intellect) can choose to go against it. Animals then, including humans, can be immoral; plants and planets cannot.[10]

Maimonides reconciled spiritual and rational life in a novel way. Early in his life, he focused on immaterial resurrection, following Ibn Sînâ. Our final blessedness must be free of body and matter in order to escape suffering and be truly eternal. When accused of denying the physical resurrection, he drew a distinction between resurrection and blessedness (*On the Resurrection of the Dead*). Citing the book of Daniel in the *Tanakh*, he said that souls and bodies will be reunited in the resurrection—at the coming of the Messiah. They will then live again in the flesh. After a second death, the blessed will live eternally and immaterially with God. This provides one of the earliest recognizable examples of nature used to exclude spiritual life—and vice versa.

The Aristotelian synthesis of philosophy and monotheism shifted life-concepts radically, separating body and mind and deemphasizing cosmic organismality. Neoplatonic interpreters of Aristotle in the fourth century started a long, gradual redefinition of Aristotelian causes and souls. Matter became material, physical stuff; souls became immaterial, non-physical stuff. Living beings became awkward hybrids of the two. Vegetable and animal souls were reimagined as faculties for biological

---

[10] This position is reminiscent of Philo's argument about the incorruptibility of stars.

activity, still awkwardly placed between the two extremes. Meanwhile, human souls were pushed firmly to the Platonic and spiritual side.

Medieval scholars took for granted a Divine order and wished to place living things within it. They began to see reason as entrenched within a physical process, even while committing to subsistent souls as the subjects of reasoning. Theologians identified spiritual and subjective life more strongly with the will, the intellect, and rational life. Nonetheless, significant tensions remained between Hellenistic and monotheistic perspectives. How can God's creation (and incarnation in Christianity) be reconciled with the eternity and/or goodness of physical matter? How does physical resurrection work? And, does it require a collaboration between vegetable and spiritual life? Thomas Aquinas began to answer these questions, developing a systematic theology that set the foundation for Christian—and non-Christian—debates in the Enlightenment and beyond.

References

Adamson, Peter. "Al-Kindi." In *Stanford Encyclopedia of Philosophy*, Spring 2015 ed. Stanford University, 1997–. https://plato.stanford.edu/archives/spr2015/entries/al-kindi/.

Al-Ghazali. *The Incoherence of the Philosophers*. Translated by Michael E. Marmura. Provo, UT: Brigham Young University Press, 2000.

Blumenthal, H.J. *Aristotle and Neoplatonism in Late Antiquity*. Ithaca, NY: Cornell University Press, 1996.

Hasse, Dag Nikolaus. "The Early Albertus Magnus and His Arabic Sources on the Theory of the Soul." *Vivarium* 46 (2008): 232–252.

Ibn Sînâ. *A Compendium on the Soul*. Translated by Edward Abbott van Dyck. Verona: Nicolo Paderno, 1906.

Ibn Sînâ. *The Canon of Medicine*. Translated by Oskar Cameron Gruner. New York: AMS Press, 1973.

Ivry, Alfred. "Arabic and Islamic Psychology and Philosophy of Mind." In *Stanford Encyclopedia of Philosophy*, Summer 2012 ed. Stanford University, 1997–. https://plato.stanford.edu/archives/sum2012/entries/arabic-islamic-mind/.

Leunisson, M. "Aristotle and Philoponus on Final Causes in Scientific Demonstrations in *APo.* II.11." In *Interpreting Aristotle's Posterior Analytics in Late Antiquity and Beyond*, edited by F.A.J. de Haas, M. Leunissen, and M. Martijn, 183–201. London: Brill, 2010.

López-Farjeat, Luis Xavier. "Al-Farabi's Psychology and Epistemology." In *Stanford Encyclopedia of Philosophy*, Spring 2016 ed. Stanford University, 1997–. https://plato.stanford.edu/archives/spr2016/entries/al-farabi-psych/.

McGinnis, Jon. "Ibn Sina's Natural Philosophy." In *Stanford Encyclopedia of Philosophy*, Fall 2016 ed. Stanford University, 1997–. https://plato.stanford.edu/archives/fall2016/entries/ibn-sina-natural/.

Maimonides, Moses. *The Guide for the Perplexed*, 2nd ed. Translated by M. Friedländer. London: Routledge and Kegan Paul, 1904–.

Maimonides, Moses. *The Eight Chapters of Maimonides on Ethics (Shemonah Perakim): A Psychological and Ethical Treatise*. Translated by Joseph I. Gorfinkle. New York: Columbia University Press, 1912.

Marder, Michael. *The Philosopher's Plant: An Intellectual Herbarium*. New York: Columbia University Press, 2014.

Martin, Raymond, and John Barresi. *The Rise and Fall of Soul and Self: An Intellectual History of Personal Identity*. New York: Columbia University Press, 2006.

Philoponus. *On Aristotle's "On the Soul" 1.1–2*. Translated by Philip J. van der Eijk. Ithaca, NY: Cornell University Press, 2005.

Simplicius. *On Aristotle's "On the Soul" 1.1–24*. Translated by J.O. Urmson. Notes by Peter Lautner. Ithaca, NY: Cornell University Press, 1995.

Sorabji, Richard. "The Ancient Commentators on Concept Formation." In *Interpreting Aristotle's Posterior Analytics in Late Antiquity and Beyond*, edited by F.A.J. de Haas, M. Leunissen, and M. Martijn, 3–26. London: Brill, 2010.

Tawara, Akihiro. "Avicenna's Denial of Life in Plants." *Arabic Sciences and Philosophy* 24 (2014): 127–138.

CHAPTER 11

Life Divided: Vegetable Life in Aquinas

In Christianity, the Aristotelian Synthesis appeared most clearly in the works of Albert the Great and his student Thomas Aquinas. Both were members of the Dominican Order and involved in the rising influence of universities, scholastic theology, and a centralized Catholic teaching authority in the thirteenth century. Aquinas crafted a new theory of life that pushed humans and vegetables farther apart. Human souls took on the subjective and spiritual life of Augustine in addition to the rational life of Plato and Aristotle. Vegetable and animal souls followed the path of Aristotle's biological activities. Commonly understood as mortal, material, or lower souls by this time, they shifted from processes in-action and in-fulfillment to proto-agents with biological faculties. The resulting awkward hybrid emphasized human transcendence and vegetable physicality but left brute animals floating in between.

ALBERT THE GREAT

Albert (1200–1280) traveled throughout Western Europe, working as an academic, priest, and bishop, but is best known for his work at the Universities of Paris and Cologne. He sought to produce accessible summaries of Aristotle, drawing on, but also critiquing Islamic interpretations. For the most part, he followed Ibn Sînâ when speaking of life-concepts. He advocated for a subsistent human soul with vegetable, animal, and rational faculties. He spoke of vegetables and animals having

© The Author(s) 2018
L. J. Mix, *Life Concepts from Aristotle to Darwin*,
https://doi.org/10.1007/978-3-319-96047-0_11

their own mortal souls acting through natural forces to bring about nutrition, reproduction, sensation, locomotion, and other life activities. The relationship of souls to bodies and the unity or plurality of the soul were both hot topics among theologians at that time.

Albert rejected universal hylomorphism, the idea that all created things are substantial only as a union of form and matter (Hasse 2008). This position had been proposed by Origen and was advanced by Ibn Gabirol, Roger Bacon, and Bonaventure. Instead he saw the soul as a distinct substance. Aristotle had spoken of the soul as both form (*eidos*) and in-fulfillment (*entelechia*). Albert called the soul substantial and distinguished it from both. A substance exists on its own, while a form must inform some matter. Nor is a soul in-fulfillment, but only possessed of a potentiality, the fulfillment of which is an actuality or activity. Thus, agents were rendered ontologically distinct from activities, and the soul was placed in the former category. Vegetable and animal souls were left in limbo; they were still described with the same potential/actual language but did not have the subsistence of human souls.

Albert embraced a compromise that proved central to late medieval life-concepts, saying, "when we consider the soul according to itself, we shall agree with Plato; but when we consider it in accordance with the animation it gives to the body, we shall agree with Aristotle" (Dales 1995, p. 98).[1] One might equally say that he used Plato for the subsistence and immortality of rational souls, but Aristotle for the material dynamism of vegetable and animal souls. He stated clearly that vegetable and animal souls arise naturally, while the rational soul is produced by God out of nothing (*Questions*, 16.10–12).

Nonetheless, Albert occasionally recognized some ambiguity. In discussing Aristotle's treatment of animals, he described plants as liminal. "[T]he genus of plants, when compared to the non-living, is living, but compared to animals is non-living" (Albert 2008, p. 227, *Questions*, 7.2). Similarly, he placed mushrooms on the border between minerals and plants, sponges between plants and animals, and children (along with "drunken and intemperate men") on the border of rationality.

Albert saw the universe as a nested hierarchy, mirrored in the feudal system (Führer 2017). God provides the first cause and final end of the cosmos. Angels and celestial intelligences, immaterial and spiritual in character, take their place directly below God. Intellectual souls, joined

---

[1] This is Dales' translation of Albert's *Summa Theologica* 2:12.69 (*Opera Omnia*, vol. 33, p. 16).

to bodies in their creation, come next. Thus, Albert rejected Plato and Ibn Rushd, for whom the body was a prison. It is good for souls to be in bodies. Below the intellectual souls rank the material things, descending through brute animals, and plants, to soulless matter. This ladder of nature or *scala naturae* is a common image through the Middle Ages and into Modernity. The system is Neoplatonic in its orientation and association of perfection with the first cause. It has lost some sense, however, of participation of all in the greater life of the cosmic soul. Individuals are distinct and moved by love of God, rather than organs of a greater life.

Drawing on Maimonides and Islamic philosophers, Albert favored reasoning from sensed particulars to universal truths, including species. Like Aristotle, he composed encyclopedic catalogs of biological observations, notably *De Vegetabilibus* and *De Animalibus*. These compendia— and the natural philosophy approach that they demonstrate—set the tone for medieval scholastic epistemology. Albert's matter-positive theology also inspired his student, Thomas Aquinas, to develop a view of life and personhood more deeply and permanently connected to physical bodies.

Thomas Aquinas

Aquinas (1225–1274) was born to wealthy parents near the town of Aquino in the kingdom of Sicily. A Dominican and scholastic theologian, he worked to reconcile the corpus of newly available Aristotelian texts with scripture and medieval theology. His most famous works were *Summa Theologiae* (*ST, Aquinas* 2006), a systematic approach to theology including natural philosophy, and *Summa Contra Gentiles* (*SCG, Aquinas* 1975), a systematic apologetics. He also wrote *Disputed Questions on the Soul* (*The Soul, Aquinas* 1951b) and a section by section commentary on Aristotle, *Sentences 'On the Soul'* (*Sentences, Aquinas* 1951a). Though initially challenged for his interest, he helped bring Aristotle back into common use among Christians and his perspectives strongly influenced Roman Catholic theology.

Hierarchical Creation

Aquinas thought in terms of a continuum—or perhaps a process of continual creation—flowing from God into matter. In this, he mirrored both Augustine and Ibn Sînâ. He stretched the continuum near to breaking, however, and substance dualism was about to break the surface.

Aquinas saw matter as a type of thing. It was not yet a substance in and of itself; it still required a form, and yet he could speak of prime matter as the potentiality of elements. Form, on the other hand, may exist without matter. Following Augustine, he divided the cosmos into visible and invisible aspects. Ontology split into two distinct categories: visible and invisible. A divided hierarchy runs from rocks to humans in the visible creation and from God to human souls in the invisible creation (*Disputed Questions*, 7). Within the invisible creation, the hierarchy descends from ideas and causes in the mind of God— Christ as Logos—through the angels, stars, and planets, to human souls. The heavens were kicked out of the "visible" creation of elements and corporeality. Though made of aether, the most rarified fifth element of Aristotle, they do not follow the rules of earthly matter. Subsequent accounts of motion were unambiguously different for angels, stars, and planets in the heavens and for vegetables and animals on earth.[2]

Aquinas considered both angels and disembodied human souls to be immaterial intelligences (*The Soul*, 7; *ST*, 1:50.1). They *can* perceive the invisible creation directly and perfectly (angels do, humans usually do not). They are not alive in the same way that animals are alive. The position appears most clearly in his treatment of stars (*ST*, 1:70.3). Because they are incorruptible, stars have no need of vegetable faculties. Because they have no elemental bodies, they lack the animal faculties. They have a principle of motion, said to the be the form of their celestial (non-elemental) body, but Aquinas does not refer to this as a soul. Angelic nature resembles the nature of human souls. Insofar as they have souls, they are exclusively soul and nothing else. The planets, meanwhile, undoubtedly have intelligence, but may or may not have souls. If they do, they rank with the angels in the invisible hierarchy (*The Soul*, 8). Citing Basil and following Philo, Aquinas stated that the plants were created before the stars to discourage idolatry (*ST*, 1:70.1).

Within the visible creation, the hierarchy ascends from minerals, through the plants and brute animals, to humans. All visible substances are hylomorphic, possessing both form and matter according to the language of the Aristotelian synthesis. The mineral world includes elements held together by basic forms and moved by basic physics or nature.

---

[2]The second account of creation in Genesis, which Augustine identified with visible, temporal creation, contains no mention of stars and planets.

In *Sentences,* Aquinas spelled out his Aristotelian psychology, setting the tone for the next five centuries of debate in biology. All living things share a common feature: souls (1:1.1). Aquinas explained the history of souls in this way. Plato sought a single soul, in which all souls participate, while the Natural Philosophers (Thales et al.) reasoned up from individuals; Aristotle sought both (1:1.12–13). From a modern vantage, Aristotle seems to side more closely with the Natural Philosophers, but Aquinas, encountering him through the mediation of *falsafah,* saw a much heavier influence of the disembodied intellect.

Aquinas described the human soul as having three parts or powers—vegetable, sensitive, and intellectual (*potentiae videlicet vegetabili, sensibili, et intellectuali*)—describing them as nested and in order of increasing perfection (1:14.199). His dual creation provided an ontological basis for distinguishing between sensation and proto-agency in the three parts.

> And these powers differ. For while the vegetative or nutritive power acts through active and passive qualities of matter, such as heat and cold and the like, the sensitive power requires no such sensible qualities for its sentient activity, though it does depend on corporeal organs; while the intellectual power acts through neither sensible qualities nor a corporeal organ, for it functions in an entirely incorporeal way. (*Sentences,* 1:14.200)

His commentary runs with the text of Aristotle's *On the Soul,* but the typology clearly follows Ibn Sînâ, perhaps going a step further toward dualism. Nonetheless, each organism has only one soul, which is neither the organism itself, nor a portion of the organism, but the organization of matter which makes it what it is. Eleonore Stump (1995) refers to it as the "configuration" of the organism, recalling more closely the dynamic Aristotelian sense of a process in-action and in-fulfillment.

In humans, as in all organisms, the vegetable faculties act through elemental physics. More specifically, the soul operates from the heart, by means of heat, acting on and through the elements (2:8.331, 348). Aquinas summed it up neatly. "Food is changed into the being of the one fed" (2:8.347). Elemental physics became the instrument of the vegetable soul as it shifts the potential substance of food from a non-living substance (i.e., flesh or mineral) to the substance of self. Because the vegetable faculties act through the passive nature of elements and heat, and because the direct object of nutrition (food) is both material and inanimate, the vegetable soul is counted as the least perfect, the lowest, and the least dignified of the souls. "Now obviously, where the thing's vitality consists entirely in growing and taking nourishment (as in plants) the vegetative principle is simply the soul or life-principle itself" (2:4.262). Plants should be identified with the vegetable faculty, itself the fullness of their soul and organization. Here, the soul is described as an active process, as in Aristotle. Aquinas expanded upon the vegetable soul with three familiar operations, listed in order of increasing perfection: nutrition, growth, and reproduction (or nutrititive, augmentative, and generative powers) (2:8.347; see also *The Soul*, 8 and *ST*, 1:78.2).

In humans, as in all animals, the sensitive faculties act on immaterial forms through material organs (*The Soul*, 13, 15). All animals receive forms by way of physical senses. Those senses are clearly localized (sight to the ears, etc.). Aquinas added internal senses, again following Ibn Sînâ: common sense (judgment in humans), memory, imagination, and estimation (particular reason and passive intellect in humans). The sensitive soul uses the vegetable soul as the vegetable soul uses the elements. All of these faculties, all of these powers, are necessarily embodied. The immaterial forms are processed in a material body. Even the particular reason is powered by the heart and localized to a region of the brain (*ST*, 1:78.4). Brute animals should be identified with the sensitive faculty, which, upheld by the vegetable faculty, expresses the fullness of their soul and organization. Aquinas expanded upon the sensitive soul with three familiar operations, listed in order of increasing perfection: sensitive, appetitive, and locomotive (*Sentences*, 1:14.201; *The Soul*, 13; *ST*, 1:78.1). Necessarily embodied, brute souls cannot survive death of the body (*ST*, 3:97.2; *The Soul*, 14; though see below for sensitive faculties in the human afterlife).

With the human soul, Aquinas made a hard move toward Platonism. In humans alone, among the visible creation, the intellectual faculty acts immaterially on immaterial forms (*Sentences*, 2:6; *The Soul*, 1; *ST*, 1:89). A nice distinction is necessary between sensible things—the forms exhibited by visible particulars, perceived fallibly through the body—and intelligible things—universal forms present in the mind of God, perceived infallibly through the intellect. Recall the forms of Plato and the archetypes of Augustine. For Aquinas they exist by themselves and yet in life, we must reach them by reasoning upward from sensible things, following Ibn Sînâ. An ontological and epistemological threshold is crossed when we move from common sense and particular reason to true intellect. God participates in an important way at this level. Thus, humans are identified with the intellectual faculty. Human souls, as intellectual agents, are made in the image and likeness of God. We are the most complex, most perfect, and most dignified living beings. Brutes and plants properly serve us as slaves.[3] Indeed, our dual nature makes us more complex and more perfect than even the angels, at least when we live into the fullness of our creation (Goetz and Taliaferro 2011, p.63).

Remnants of Aristotelian intellect remained. In this life, human intellect uses and, critically, is dependent upon both sensitive and vegetable faculties. Damage to physical organs can irreparably harm the rational soul's ability to understand itself or anything else (*On Spiritual Creatures*, 2, Aquinas 1949). In the life to come, human souls pull some aspects of sensitive (but not vegetable) faculties with them. This attempt to blend Aristotelian and Platonic intellect ends up physicalizing the vegetable faculties (and plants) while spiritualizing the sensitive faculties. Three areas highlight the awkwardness of this tack: birth, will, and the afterlife.

Multiple Births

In the invisible creation, God made the species and forms of plants. Aquinas spoke of plants having a hidden life, created on the third day, but not yet materially manifest (*ST*, 1:69.2). The processes of nutrition and procreation are etiologically prior to their action in the world and they persist invisibly in the mind of God. This idealization of typological processes and species pushed fundamental areas of biology out of the realm of sensible things, creating an epistemological distance and, in the mechanical philosophy, an epistemological barrier to biological knowledge. Species and individuality, under the head of formal causes, and organic purpose, under the head of final causes, became removed from the empirical world. Only discrete individual forms, propagated through visible seed and active within time, are available to the bodily senses. The forms of species are not. Aquinas and Ibn Rushd thought humans could process those sensible forms and, thus, discern intelligible forms. Later philosophers were more skeptical.

The stress points appear with spontaneous generation and human embryos. Vegetable souls are clearly mortal and material. In the normal

---

[3] *ST* 2(2):64.1, citing Augustine's *City of God* 1:20 (Augustine 1950). It is worth noting that Aquinas here suggested that humans move themselves, while brutes and plants are moved by external reason (i.e., God).

course of life, they physically pass from generation to generation: traducianism. A small piece of the paternal soul breaks off and, through inherent heat, sparks a soul in the offspring. Spontaneous generation is also possible (*Sentences*, 2:7.313). Without a paternal soul, the new organism depends on environmental heat and planetary motion to quicken. The explanation, implausible to modern ears, represented lingering aspects of Neoplatonism and cosmic organism.

Human reproduction provided a harder test. Along with most thinkers in the Aristotelian lineage, Aquinas struggled to make the human soul both a unity, embracing vegetable, sensitive, and rational faculties, and a peculiarity, uniquely immaterial and dignified (*ST*, 1:118.2, reply 2). He noted that human embryos lack reason. They must, at first, be motivated by a vegetable soul alone. When, sometime later, they acquire sensation and then reason, they require sensitive and rational souls. And yet, there can be only one soul at a time. The early soul cannot be rational without exercising reason because (again following Aristotle) the rational soul only exists as an active process; potentiality is not enough. Nor can Aquinas make sense of a vegetable soul becoming rational. He proposes, therefore, that human embryos begin with a vegetable soul, traduced through the heat of the paternal sperm. When development reaches a certain level, the vegetable soul is annihilated and replaced by a sensitive soul, also kindled from the heat of the sperm. God then specially creates the rational soul and places it within the already organized body, annihilating the sensitive soul. In a way, there are three births and two deaths before parturition.

Will and Intellect

The appetite, characteristic of all animal souls, takes on a special character in humans. Humans have more freedom in their appetites, allowing them to choose what they will pursue. Aquinas called this the intellectual appetite or the will (*ST*, 1:82.5). Spiritual life depends upon a will that, through the intellect, has the power to choose rightly. The intellect of itself perceives the good clearly, but only the will, combining sensitive and rational faculties, can choose good or evil (1:82.3). As a sensitive faculty, it works through the heart (*SCG*, 2:72; *On Spiritual Creatures*, 3 ad. 4). This places human spiritual life firmly in the realm of the material, sensitive soul, though it must be aided by the intellect. So, too, resurrection life pulls the sensitive soul past death. "The sentient [or sensitive] soul in brute animals is corruptible, but since in man the sentient soul is identical in substance with the rational soul, it is incorruptible" (*The Soul*, 14, reply 12).

Resurrection

Aquinas found himself in a bind. Plato, Plotinus, Ibn Sînâ, and Maimonides had rejected the body in eternal life. True freedom lay only in immateriality. Aquinas wanted to affirm physical embodiment as good in creation, both proper and eternal in resurrection. He embraced a modified Aristotelian hylomorphism. Humans have a subsistent soul, which is nonetheless essentially embodied. It can be alienated from the body, but it is incomplete when this happens.

The soul separates from the body in death. Each human soul is individually created by God for a specific body and, though it can subsist, it does so as a partial being. The disembodied dead are closer to the shades of Homer than the freed star-stuff of Plato. "My soul is not I" (*Commentary to 'I Cor 15:17–19,'* cited in Still 2002, p. 108). At death, the soul departs from the body and exists on its own. Having committed to sensitive faculties as necessary for knowledge—and having committed in principle to the mortal physicality of sensitive souls—Aquinas must somehow explain sensation in disembodied souls. This topic takes up the last eight of twenty-one articles in *The Soul* (and *ST*, 1:89). Echoing Plato, Aquinas finds it fitting that the those who chose material things in life should be punished by material things in death. Unrighteous souls suffer from fire in Hades; their sensitive faculties persist.

The third book of *Summa Theologiae* spells out the final ends of human life. Incorporeal existence (in Hades) cannot express the fullness of human life. "Hence no soul will remain forever separated from the body. Therefore, it is necessary for all, as well as for one, to rise again" (3:71.2). When the world ends and the heavens cease their motions, all the dead will once again take on bodies (3:77.1). They will not take on vegetable and animal souls, as those are passible and subject to change. Immortal beings need neither nutrition nor reproduction (3:81.4). Aquinas never clearly explained how they might have physical bodies which are, nonetheless, unchanging. The blessed elect move on to Paradise, where they are beyond suffering; they are literally impassible, incapable of being acted upon (3:86). The damned move on to Hell proper, where their bodies are also physically impassible; they exist forever, though they are sensibly subject to pain. The details remain unclear. Aquinas rejected both Platonic idealism and the dynamic hylomorphism of Aristotle. On this question, he did not provide a clear alternative.

HIERARCHY

At first glance, Thomas Aquinas represents a triumph of Parmenides over Heraclitus. The cosmos exists as a fixed hierarchy running from God to humans and from humans to minerals. Following Plato, the hierarchy is continuous, but following Neoplatonic Aristotelianism every level reflects a stable eternal form or archetype. Forms had become idealized and immortalized in the mind of God without recourse to the organicism and progress of Pythagoras and Plato. Meanwhile, physical matter had become alienated from spirit. Aquinas held on to the last remnants of an organic cosmos, believing that all things are motivated by the love and will of God (an external agency of which they are patients). His successors lacked even this. Moving to mechanical metaphors, they firmly distinguished physical machines from the mental/spiritual agents that motivate them and give them purpose.

In this context, hybridity became heresy. Boundary crossing, whether between form and matter, spiritual and physical, male and female, or psychological and physiological threatened the perfect order, despite the inherent liminality and flow of biology. The line between brute animal and human was not new, but it became ontological and irreconcilable in the wake of Aquinas. In humans, spiritual souls move physical bodies. Animals and plants are only physical bodies, moved by Divine cosmic laws.

To make sense of the human dichotomy, the sensitive soul became a way station, a place where vegetable mechanism met psychic experience. Sensation developed a front end, with stimuli impacting the body, and a back end, with the body passing the messages on to an immaterial soul. Locomotion or willed motion likewise had a back end, with psychic appetite signaling the body, and a front end, with the body carrying out motions.

Vegetation, still characterized by nutrition, growth, and reproduction, became the type for physical, natural life; subjective interiority, characterized by intellect and resurrection, became the type for ensouled, spiritual life. Sensitivity and will, ever awkward as they participate in both, remained problematic. They were increasingly pushed toward the vegetal in brutes and toward the spiritual in humans. And yet plants and brutes retained an organismality and proto-agency based on their eternal natures. How can we know anything about them, without access to that immaterial world? Working out that conundrum set the groundwork for modern science and modern biology.

Modern readers, especially those versed in the natural sciences, may have a justified skepticism about Aquinas' philosophy of nature. Most familiar will be the divide between visible and invisible (now often thought of as natural and supernatural), his appeal to formal and final causes in the mind of God, and his clear ontological distinction between humans and other animals. I would not wish to minimize the problems inherent in those positions, especially in biology. Nor would I wish to minimize the insights of the scientific revolution. But, the Thomist position was itself a response to a long chain of arguments about biology. As we turn to the scientific revolution, and debates more recognizable as "biology" in the modern sense, it will be important to recognize that history. The "orthodoxies" being overturned in the seventeenth century were, in many cases, only a few centuries old. Again and again, scholars proposed continuities of earth and heavens, matter and life, life and reason. What changed was *how* they were seen to be continuous and the ways we explain their properties.

From their rediscovery around 1200 until the seventeenth century, "Aristotelian" perspectives dominated the epistemology and natural philosophy of Europe. This synthesis of Neoplatonism and monotheism (predominantly Christianity) set the language and research agenda for countless explorations of living things. They were perceived in a dual hierarchy, running from rock to humans and from humans to God, tied together by their common participation in God's will and order. In Greek terms, that order was the *Logos* of the *Cosmos*. In Christian terms, it was Christ as Word of God. The High Middle Ages and Renaissance (roughly 1000–1500) embraced the *scala naturae*, linking all reality in a harmonious pyramid pointed upward toward God. The ideal universe matched the ideal kingdom, with a single ruler, a few courtiers, and many subjects, perfectly ordered and arranged. In this way, each aspect of reality could be interrogated to provide a better understanding of the whole.

Plato, because of his cosmic organism metaphor, saw organismic principles as essential to understanding reality. Aristotle, likewise, used final causes, as the end and organizing principle of living things. By the Middle Ages, this teleology had shifted to individual intention, a prospective will by which agents cause change in the world. Vegetable, animal, and rational (proto-)agency could all be explained by movement toward God, while God's agency drew the world upward. These two processes were joined in final causes: biological action and divine intention.

References

Albert the Great. *B. Alberti Magni … Opera Omnia*, 38 vols. Edited by S.C.A. Borgnet. Paris: Vives, 1890–1899.

Albert the Great. *Questions Concerning Aristotle's "On Animals."* Translated by Irven M. Resnick and Kenneth M. Kitchell, Jr. Washington, DC: Catholic University of America Press, 2008.

Aquinas, Thomas. *On Spiritual Creatures*. Translated by Mary C. FitzPatrick and John J. Wellmuth. Milwaukee: Marquette University Press, 1949.

Aquinas, Thomas. *Sentencia Libri "De Anima."* With English translation by Kenelm Foster, O.P., and Sylvester Humphries, O.P. New Haven: Yale University Press, 1951a. HTML edition by Joseph Kenny, O.P. Online at http://dhspriory.org/thomas/DeAnima.htm#24 [*Sentences*].

Aquinas, Thomas. *The Soul*. Translated by John Patrick Rowan. Eugene, OR: Wipf and Stock, 1951b.

Aquinas, Thomas. *Summa Contra Gentiles*. Translated by Anton C. Pegis. Notre Dame, IN: University of Notre Dame Press, 1975 [*SCG*].

Aquinas, Thomas. *Summa Theologiae*, 61 vol. New York: Cambridge University Press, 2006 [*ST*].

Augustine. *The City of God*. Translated by Marcus Dods. New York: Random House, 1950.

Dales, Richard C. *The Problem of the Rational Soul in the Thirteenth Century*. New York: E. J. Brill, 1995.

Führer, Markus. "Albert the Great." In *Stanford Encyclopedia of Philosophy*, Fall 2017 ed. Stanford University, 1997–. https://plato.stanford.edu/archives/fall2017/entries/albert-great/.

Goetz, Stewart, and Charles Taliaferro. *A Brief History of the Soul*. Malden, MA: Wiley-Blackwell, 2011.

Hasse, Dag Nikolaus. "The Early Albertus Magnus and His Arabic Sources on the Theory of the Soul." *Vivarium* 46 (2008): 232–252.

Still, Carl N. "Do We Know All After Death? Thomas Aquinas on the Disembodied Soul's Knowledge." *Proceedings of the American Catholic Philosophical Association* 75 (2002): 107–119.

Stump, Eleonore. "Non-Cartesian Substance Dualism and Materialism Without Reductionism." *Faith and Philosophy* 12 (1995): 505–531.

PART III

Death

CHAPTER 12

Mechanism Displaces the Soul

After Aquinas, the Aristotelian concept of souls, carefully tended for two millennia, started to unravel. Renaissance humanism and skepticism challenged the reliability of ancient authorities and the transcendence of the intellect. Protestant theology changed both method and focus in scriptural interpretation. New observations and better communication broadened thinking about life. By the end of the seventeenth century, the mechanical philosophy had begun to take the place of the Aristotelian synthesis. The new perspective placed formal and final causes beyond the scope of natural sciences. For Descartes, this involved full-blown ontological dualism. For Gassendi, it meant epistemic humility. In either case, an interesting question arises: where did the formal and final causes go? Nutrition and reproduction, along with curtailed versions of sensation and locomotion received new, mechanical accounts. Will and intellect, on the other hand, moved to a more rarified, and ontologically distinct, immaterial realm of thought. Displaced from nature, they made both human and Divine agency unnatural—artificial, supernatural, or both.

Physiological and psychological accounts of life took on distinct and irreconcilable vocabularies, with the term "soul" frequently reserved for the latter. Physics and physiology began to refer exclusively to mechanical accounts of matter in motion. Empirical reasoning, based on induction from observation, proved extremely effective in explaining physical, chemical, and eventually biological phenomena. Meanwhile psychology

L. J. Mix, *Life Concepts from Aristotle to Darwin*,
https://doi.org/10.1007/978-3-319-96047-0_12

143

and the human sciences focused increasingly on subjectivity. The two languages drifted apart.

The increasing availability of Greek texts allowed for deeper examination of traditional authorities. In the twelfth and thirteenth centuries, increased trade had brought classical works into Western Europe. For the first time, scholars could read Aristotle and Plato in the original Greek. Jewish and Christian scriptures similarly began to appear in their original languages, encouraging multiple translations. The works of Galen and Lucretius provided alternatives to the Aristotelian/Neoplatonic orthodoxy. All these movements were accelerated by increased prosperity and the introduction of a printing press with movable type in the fifteenth century. Though still a luxury item, books became more readily available.

WILLIAM OF OCKHAM

A contemporary of Aquinas, William of Ockham (c. 1287–1347) was a Franciscan philosopher born in England. Called to the Papal Court in Avignon in 1324, he spent the last 20 years of his life focused on questions of poverty within the church and related political affairs. Prior to that time, his work as a writer and teacher set the groundwork for an alternative to the Aristotelian synthesis. Two ideas proved central to the development of life-concepts: nominalism and voluntarism. Both were essential to medieval reasoning about souls, seeds, and species.

Plato or Aristotle? Philosophers return again and again to one question. Is the world extended downward from ideal being or built up from base material? Ockham cemented the move toward Aristotle begun in Islamic *falsafah*. Aquinas' lingering Platonism included *Platonic realism*, wherein universal ideas, predicated of many things (e.g., white, horse, plant) are more real than their particular instantiations. Ockham held the contrary position of *nominalism*. Universals are simply names we use to describe collections of properties. Particular things are real; universals are no more than the groups we put them in.

Margaret Osler (1994, 2010) describes a second argument, related to will and intellect. The attribution of these two faculties to God shaped epistemological reasoning from the Late Middle Ages to the Enlightenment. Theologians, describing the dual creation, contrasted two types of Divine agency. Aquinas argued that God's intellect and the invisible creation correspond to God's *absolute power*. God establishes rules in eternity. These rules bind God as much as they bind creation. Thus, God's will and agency (in time) are constrained by God's intellect, a position called *intellectualism*. Human intellect, by participation, may know the Divine Law that governs reality.

Ockham denied that God is so limited—or perhaps that humans may be so informed. He emphasized the complete freedom of God's will: *voluntarism*. Divine Law does not constrain Divine agency. Voluntarism allowed miracles, but it also required empiricism. Without necessary and eternal truths in the mind of God—and human access to them via the intellect—we must reason upward from observed regularity to general rules. This would not have been possible without the work of Ibn Sînâ, Maimonides, Ibn Rushd, and Albert on tools for induction.[1]

This new form of skepticism both challenged universal laws and supported universal regularity. It provided a cornerstone for Renaissance thinking as ancient authorities and a priori intellectual knowledge were tested against concrete, systematic observations. The new philosophy emphasized God's will and the role of individual human wills to understand it. It made empiricism and modern science possible.

The Protestant Reformation

Christian Reformers also shaped biology, mostly through their rejection of allegorical interpretations of the Bible. Their flattened epistemology limited speculation on the soul. In his commentaries on Genesis, Martin Luther (1483–1586) asserted that God intends only to teach us about the visible creation. Augustine's allegories should be avoided in favor of a more literal reading (*Luther's Works*, 1:4–5, Luther 1955–1986). The dual creation disappears from most Protestant thought. Citing Lucretius, Luther questioned our ability to have any definite knowledge of the soul. He thought we should focus on what God commands.

Luther's comments on plants were minimal and flavored by contemporary science (1:36–54). He referred to plants as "kitchen and provisions" for humanity. They have continuity of type though Divine foresight, not created archetype. He believed that the term "bearing seed" in Genesis 1:11–12 provided a clear distinction between plants and animals. Plants have a seed phase; animals propagate from body to body.[2] This worked against the idea of animals as advanced plants.

---

[1] The intellectualist-voluntarist debate echoed the debate between Ibn Sînâ, defending eternal truths, and al-Ghazâlî, defending occasionalism and the possibility of miracles.

[2] By modern standards, it is unclear how seed and egg differ; nor do all plants have proper seeds.

Luther also emphasized the dramatic curse of the Fall, which impaired human will and intellect. Prior to the Fall, Adam could look at things and perfectly know their nature, allowing him to name them properly (1:120). After the Fall, knowledge requires observation. This created an important divide between Catholic and Protestant epistemology regarding living things. Catholics trust the intellect to perceive God, but consider it obscured by both the body and the fallen will. Protestants consider the intellect fallen in a way that cannot be purified unless the will (moved by the Holy Spirit) chooses God. The two approaches shape how and when they are willing to be convinced by empirical versus scriptural arguments.

Luther cited Aristotle in saying that souls motivate and organize bodies in all animals (8:14–15). Human souls are unique. They bear two kinds of life, physical and immortal, because they are in the image of God (1:56–57). This insures life after death. He restated Augustine's ambivalence about how human souls come to be. The souls of brutes presumably pass by traducianism. Human souls may pass the same way but may also be specially created (8:91). Planets, meanwhile, are driven in their orbits by God directly, needing no special soul nor intelligence to move them (1:29–33).

John Calvin (1509–1564) made even stronger claims about the effects of the Fall. He cited Plato on the immortality of the soul and the soul's three faculties: intellect, irascibility, and concupiscence (*Institutes*, 1:15.6–8, 5.5, Calvin 1997). He called this latter division true, or at least plausible, but not useful. With Luther, he believed that the will and the intellect are the only aspects of the soul worth studying. Prior to the Fall, both were perfect, but now they are both so corrupt that they are incapable of choosing or perceiving the Good without Divine intervention. This leads to an almost total dependence on scripture for epistemology. The world was made in six days and has lasted for roughly 6000 years (1:14.2). This position was softened somewhat by a later claim that the will is placed between sense and reason (2:2.2, 12–13). Humans retain intelligence regarding natural events and a love for truth, but lack intelligence regarding the supernatural (God and ethics) and the proper object of truth. Natural science remains practical, so long as it is properly cordoned off from God, faith, and morals.

The invisible creation still lurks within some passages. "[E]ach species of created objects is moved by a secret instinct of nature, as if they obeyed the eternal command of God, and spontaneously followed the course which God at first appointed" (1:16.4). Here, special providence, God's perfect ordering of each event individually, was paired with the more traditional idea of inherent orderliness.

THE MECHANICAL PHILOSOPHY

The greatest change in the vegetable life-concept arose as a result of the *mechanical philosophy*. Early seventeenth century philosophers, led by Pierre Gassendi and René Descartes, began to speak of the physical world as a cosmic mechanism and of living things as automata. Both reduced natural accounts to matter in motion, eliminating the use of final and formal causes. While both pushed final causes out of the natural world, neither eliminated them completely. Nor did they stop using the language of souls. The mechanical philosophy can be viewed in at least three ways: the machine metaphor, ontological elimination, and etiological reduction. Although many philosophers employed more than one, they make distinct claims.

The Machine Metaphor

At the surface level, the mechanical philosophy reflects a shift in metaphors. Prior to the seventeenth century biological organization was the dominant image. Non-living things were exceptions to the rule. They lacked the *telos* or finality of fuller participation in the cosmic life. Aristotelian *mechanics* described that small set of bodies that had no final causes of their own. Humans craft machines and give our purposes to them. Thus, their ends are extrinsic and their function can be described by appeal to natural laws plus human design and motivation. In other words, they require a designer and a source of energy.

Georges Canguilhem pointed out that "the construction of a machine can be *understood* neither without purpose nor without man" (Canguilhem 2008, p. 86, from "Machine and Organism," italics in the original). The machine metaphor requires the pieces to fit together according to a plan.

> It seems to us, then, that it is an illusion to think that purpose can be expelled from the organism by comparing it to a composite of automatisms, no matter how complex. So long as the construction of the machine is not a function of the machine itself, so long as the totality of

an organism is not equivalent to the sum of its parts (parts discovered by analysis once the organism has already been given), it seems legitimate to hold that biological organization must necessarily precede the existence and meaning of mechanical constructions. (ibid., p. 91)

In other words, the machine is a prosthetic, extending human intentions into an artificially organized collection of elements (Mix 2014). The founders of the mechanical philosophy had no intention of eliminating final causes. They simply transferred them from nature to God as creator and humans as agents within the creation.

The machine metaphor was not the only—nor even, I would argue, the most important—aspect of the mechanical philosophy. It was, however, highly influential in the development of life-concepts, as philosophers asked whether vegetables and animals are agents or automata. It suggested questions about how machines are designed and powered. Mechanists in the seventeenth century could defend their Christian credentials. How could they be atheists if their theory of a clockwork universe required both a clockmaker and someone to wind the spring? Thus, the design argument—for the universe and for individual organisms—fell almost necessarily out of the machine metaphor. It is worth noting, however, that Canguilhem's quote provides an antidote. It would be precisely by making "construction of the machine" into "a function of the machine itself" that Kant and Darwin transcended the machine metaphor in later centuries.

Michael Ruse provides an excellent history of the move from organic to mechanical metaphors in *Science and Spirituality* (2010), and a lengthier history of the design argument in *Darwin and Design* (2003). Evelyn Fox Keller looks at the rise and fall of the machine metaphor in developmental biology in *Making Sense of Life* (2002). Much ink is spilt on whether mechanical metaphors are sufficient for biology and to what extent such metaphors entail a designer. It is not my intention to resolve those arguments here. I only wish to say that the metaphor was important to seventeenth century biology and, at the time, was popularly coupled with an appeal to design. Other aspects of the mechanical philosophy were more enduring and more important for biology.

Ontological Elimination

A few seventeenth century philosophers did try to eliminate all final causes and, with them, any hint of an immaterial or immortal soul.

Epicurus and Lucretius, whose works provided inspiration for the new worldview, had made just such claims in their materialism, along with a strong conviction that gods do not interfere with the natural order. Pietro Pompanazzi (1462–1525) and Thomas Hobbes (1588–1679) both embraced the mechanical philosophy in a way that called for strict materialism, including a fully mortal material soul in humans (Osler 2010, p. 145).[3] Two centuries later, Thomas Henry Huxley (1825–1895) embraced precisely this kind of elimination in biology. By adopting Epicurean principles, the mechanical philosophers became identified with ontological elimination and the expressly atheist sensibilities of these authors. It may be precisely this tension that caused others to so clearly appeal to supernatural causes and the design implications of the machine metaphor (Osler 2010, p. 89).

Etiological Reduction

A more modest proposal involved simplifying natural accounts. Aristotle had listed four avenues of explanation: material, efficient, formal, and final causes. The mechanical philosophers embraced only the first two. They further limited explanations by insisting the material in question must be, ultimately, the basic units of physical being. These building blocks of the natural world had only a small number of *primary qualities* that explain all of their properties. The more familiar and diverse properties encountered by humans, such as color, smell, etc., they called *secondary qualities*.

The etiological reduction does not require a machine metaphor. Many views of matter in motion do not involve automata or designers. Twentieth century "mechanists" embrace probabilistic and information-based metaphors. Both work without formal and final causes and without the language of clockwork. Nor does an etiological reduction require the complete elimination of formal and final causes. Descartes and Gassendi displace them, claiming that they are inappropriate when explaining physical and physiological events, but still important for psychology and theology. Some of reality is, for them, not natural. By so clearly defining a natural epistemology, the mechanical philosophers necessitated a category of super-nature to contain things that do not fit.

---

[3] See also Martin and Barresi (2006); Pompanazzi, following Alexander of Aphrodesias, argued that the human soul may be accidentally immortal, but is not so essentially.

Vegetable, animal, and rational life stubbornly refused to sit neatly in natural accounts for precisely the reasons articulated in antiquity. How do we understand the unity, motivation, and organization of living things? How do we understand their unpredictable motion? Nutrition and reproduction seem almost amenable to matter in motion accounts. Intellect appears intractable. Sensation and varying levels of agency fall somewhere in between. The forced divide in epistemology in the seventeenth century had the same effect as the forced divide in ontology in the thirteenth. Philosophers and biologists awkwardly scrambled to force aspects of life into one category or the other. From here on, I will use the term mechanical philosophy to refer only to the etiological reduction. I focus on the boundaries of nature and empiricism as they are set by particular authors.

PIERRE GASSENDI

Pierre Gassendi (1592–1655) built on Ockham's voluntarism to propose a new science of the natural. Born in southwestern France, he was a Roman Catholic priest with a position at the cathedral in Digne. In addition to teaching philosophy and theology, he conducted experiments and joined the *Collège Royal*. Gassendi attempted to reconcile Christianity with Epicurean philosophy (Osler 1994, pp. 36–77). He reduced all accounts (of the physical world) to efficient causes acting on atoms. These atoms he understood using only the primary qualities of magnitude, figure, and heaviness. He exported all final causes to the mind of God, operating like quasi-efficient causes, as in Aquinas. They are external motivators. They lie outside the things motivated but also beyond the ken of natural science. With a humble approach to natural science, Gassendi believed it provides only probable and limited knowledge of reality. Some things, including final causes, exist in the same realm, but beyond our ability to know through sense experience.

Drawing heavily on Lucretius' *De Rerum Natura*, Gassendi proposed a cosmos made up of atoms moving through the void. Unlike Lucretius, he saw God actively participating. The presence of order requires an orderer and the quality of the order reveals God's character. With Lucretius, he divided the human soul into *animus* and *anima*. The Platonic *animus* accounts for human reason and immortality. Its existence defines one of the borders of mechanical empiricism, because it exercises agency through the human body. It operates the machine. Gassendi placed it in the head.

Every living body, including that of humans, bears a material, mortal *anima* that organizes and activates it. Made of rarified atoms, the *anima* provides vital heat and accounts for the traditional vegetable and sensitive faculties. God created the first *animae* in creation and they pass from generation to generation with one *anima* igniting another through semen, just as one candle lights another. Nutrition fuels the flame. These mortal souls exist within the natural world.

RENÉ DESCARTES

René Descartes (1596–1650) set the groundwork for modern science and philosophy by providing arguments for both a priori and empirical knowledge claims. Born in France, he traveled around Europe as a self-funded philosopher. Descartes inaugurated an ontological and epistemological paradigm that remains highly influential into the twenty-first century. Many historians credit him as the father of modern philosophy and the turning point from medieval to modern science.

Beginning with his famous *Cogito* argument (I think; therefore, I am), Descartes aspired to universal knowledge. His approach begins with internal reflection, a school of epistemology that comes to be known as *rationalism*. Reason alone, without input from the senses, may know truth. With the collapse of Neoplatonism, the rational soul no longer participates directly in the mind of God, ruling out that avenue to perfect knowledge. Descartes was unsatisfied with the probable knowledge of Gassendi, however. Looking to Aquinas and the intellectualists, he appealed to necessary truths. These can be known in a general way from reflection alone or about specific objects by observation of particulars (using the methods developed in the Aristotelian synthesis).

Descartes argued that God created the world by absolute power, but maintains it by ordered power (Osler 1994, pp. 130–148). God never, under any circumstances, interferes with the laws set down in creation, thus there are no miracles (in the Enlightenment sense of divine intervention contrary to natural law). For this reason, Gassendi, Boyle, and Newton all critiqued him for going too far with intellectualism. For his part, Descartes complained about Gassendi's use of final causes. He said they were nothing more than Divine efficient causes, God inelegantly tinkering with the clockwork. A perfect engineer would have gotten it right the first time (ibid., p. 162). Descartes thought that Gassendi was attempting to peer into the mind of God when he spoke of God's intentions. Thus, Descartes had a much stricter position than Gassendi, closer to an ontological elimination. He thought that efficient causes are sufficient to all explanations in the physical world, once again excepting divine and human agency.

Cartesian Biology

Descartes marks a difficult transition in life-concepts because his ontology and epistemology were so completely integrated. He provided a science of the natural world by self-consciously extracting God, souls, and final causes from it and placing them in a separate world. He sought to collapse vegetable and animal life into physics, while promoting rational, subjective, and spiritual life into a non-physical mind. Unlike Gassendi and proponents of the medieval syntheses, he denied any hidden or invisible order within his accounts of physical nature. No final causes and no essences; just matter in motion. But, if there is no order inherent within the world, how can we claim knowledge about it? Both the order and observer must be outside. He used the rational life-concept to posit a cosmic order in the mind of God. He added human minds that can perceive it through intellect. Subjective life in humans lets us abstract the scientific observer from the system being observed. The very order that allows us to attribute unity, function, and organization—necessary for vegetable and animal life-concepts—becomes extrinsic to the physical systems they describe. For Descartes, this division was intuitively obvious and useful. For later thinkers, it would prove problematic.

Descartes divided the world into two types of substance: *res cogitans* or thinking things and *res extensa*, things that are extended in space. The two categories were mutually exclusive. The *res cogitans* are indivisible rational agents, which he called *mens* (mind) or occasionally *anima* (soul). The *res extensa* include all manner of things that take up space. Unlike Gassendi, he believed in neither atoms nor void. Matter is infinitely divisible and fills the entirety of the universe. The only primary quality of matter is extension. All other physical properties flow from this.

In part five of the *Discourse on Method*, Descartes set forth his mechanization of animal life, including the human body. Animals are merely heat-driven hydraulic (or pneumatic) machines (Voss 2000). Following the Stoics and Galen, he invoked three types of breath or material spirits instrumental in bodily motion. The less rarified spirits exist as blood fractions, but the third, the *animal spirits*, are fine enough to pass through the artery walls into the brain. More rarified than air, but less than flame, animal spirits pass through hollow nerves to actuate sensation and motion (Osler 1994, p. 220).[4] This provides an animal life-concept that is physical, mortal, and in line with the mechanical metaphor. No mention was made of plants, but in collapsing animal life into mechanical physics, it seems reasonable that plant life would be included.

In the *Treatise on Man*, Descartes claimed that he had done away with any need for vegetable and animal souls

> [I]t is not necessary to conceive of this machine as having any vegetative or sensitive soul or other principle of movement and life, apart from its blood and its spirits, which are agitated by the heat of the fire burning continuously in its heart—a fire which has the same nature as all the fires that occur in inanimate bodies. (Descartes 1984, 1:108)

He was confident asserting mechanisms for nutrition and reproduction as well as a front end for sensation and willed motion. Following tradition, sensory input occurs in the body along with common sense, imagination, and memory. In humans alone, these faculties will be tied into a back-end operator: the human mind.

The mechanization of life required three difficult elements: design, power, and operation. The organization present within organisms (the ability of the pieces to coordinate mechanically), comes from God, according to Descartes. That the pieces fit together clearly indicates some designer. Why they fit together remained obscure, even to him. Nonetheless, he complained that Gassendi was both inelegant and overambitious in his use of the final cause (Osler 1994, p. 162). He was inelegant in using final causes as divinely motivated efficient causes. He was overambitious in thinking we may know the mind of God. Better to have God design the universe initially and avoid subsequent meddling. The power for organisms also comes from the first creation, just as in Gassendi. Operation also presented challenges. The *res cogitans* human mind steers the *res extensa* human body and receives signals from it by

---

[4] "Animal spirits" is the Latin cognate of soul-breath or *pneuma psychikon*; this was the middle of the three breaths in Stoic biology.

means of animal spirits in the pineal gland (Lokhorst 2017). It remains unclear exactly how Descartes intended to bridge the divide at that location and the idea was ridiculed by later authors.

With no back-end operator, animals have very limited function. Like a thermostat, they can sense but cannot prefer one sensation over another. They have neither pain nor pleasure.[5] The animal receives an objective signal input and responds automatically, without a subjective experience of being helped or harmed. Nor do animals have an efficient cause operator as humans do; their actions follow automatically from divine action without appeal to an additional will (Descartes 2017, pp. 119–120).

Humans achieve numerical identity (they are the same person through time) only through the association of an indivisible mind. Animals and plants lack this method of differentiation making the idea of biological individuals and species at best arbitrary and at worst nonsensical. Their boundaries are not objectively verifiable. They are all material flux with nothing to unite them across time and space.

The mind/soul, a spiritual and subjective life-concept, must be removed from the natural world in order to make the world mechanical. The human mind cannot work without formal causes to define it and final causes to drive it. And yet, it must perceive real, natural particulars if it is to hold knowledge about nature. The details of that connection remained problematic for centuries, nowhere more dramatically than in biology.

LIFE IN SPACE

I cannot pass on without adding a brief note about conceptions of life beyond Earth. Throughout history, people have wondered about the existence of life elsewhere (Crowe 2008). In some eras, philosophers invoked both God's abundance and infinite, purposeless variety in defense of a populated cosmos. Other eras found the idea ridiculous based on human dignity or contemporary physics.

The debate shifted during the Renaissance, as life-concepts moved from Aristotelian and Neoplatonic to mechanical. The positions

---

[5]The position follows from *Discourse on Method* 5 and is defended by Descartes in response to critiques; see Descartes (2017, pp. 102–105, 150–151). Lennon (2000) presents arguments that later in life Descartes may have softened his position.

remained the same, but the language changed. Before this time, philosophers spoke of a "plurality of worlds." The stars and planets are living beings within the same organic cosmos as humanity. Other organic cosmoses may also exist. In modern language, we might compare this to multiverse theories or the existence of other light-cones within one physical universe. After the Renaissance and after the Copernican revolution, the stars and planets were no longer treated as alive. The question became whether the stars are distant suns, as with Giordano Bruno (1548–1600), and whether the other planets are inhabited, as with Bernard le Bovier de Fontanelle (1657–1757) and Christiaan Huygens (1629–1695). Planets ceased to be animals and started to be locations where animals might dwell.

During the Renaissance, philosophers and theologians began to look at the world in a radically different way. The organismic cosmos of Neoplatonism disappeared, taking with it confidence in intellectual perception and the carefully crafted bridge running from God to matter. In its place, a mechanical universe arose, thanks in large part to Gassendi and Descartes. This new universe had no place for the gradations of life, agency, and value present in the old system. Physics runs through all of nature, including vegetable and animal life, even in humans. Subjective and spiritual life were cordoned off, beyond the reach of natural science for Gassendi or placed in a whole new realm by Descartes.

The mechanical philosophy did not catch on in all the sciences immediately. It revolutionized physics with the work of Galileo Galilei (1564–1642) and Isaac Newton (1643–1727). In chemistry, it competed with the chemical philosophy for two more centuries. Robert Boyle (1627–1691) set the stage for a mechanical chemistry, but it did not dominate until the time of Antoine Lavoisier (1743–1794) and John Dalton (1766–1844). Life proved even more challenging as biologists tried to account for the organization and apparent purpose of living things. This led to skepticism and sometimes outright denial of mechanical biology throughout the eighteenth and nineteenth centuries. Only with Charles Darwin (1809–1882) and Gregor Mendel (1822–1844) did mechanical explanations begin to gain traction.

REFERENCES

Calvin, John. *Institutes of the Christian Religion*. Translated by Henry Beveridge. Grand Rapids, MI: Eerdmans Publishing, 1997.

Canguilhem, Georges. *Knowledge of Life*. New York: Fordham University Press, 2008.

Crowe, Michael J., ed. *The Extraterrestrial Life Debate: Antiquity to 1915, A Source Book*. Notre Dame, IN: University of Notre Dame Press, 2008.

Descartes, Rene. *The Philosophical Writings of Descartes*, 2 vols. Edited by J. Cottingham, R. Stoothoff, and D. Murdoch. Cambridge, UK: Cambridge University Press, 1984.

Descartes, Rene. *Discourse on Method and Meditations on First Philosophy*. Edited by David Weissman. Rethinking the Western Tradition. New Haven: Yale University Press, 1996.

Descartes, Rene. *Meditations on First Philosophy with Selections from the Objections and Replies*, 2nd ed. Cambridge Texts in the History of Philosophy. New York: Cambridge University Press, 2017.

Keller, Evelyn Fox. *Making Sense of Life: Explaining Biological Development with Models, Metaphors, and Machines*. Cambridge, MA: Harvard University Press, 2002.

Lennon, Thomas M. "Bayle and Late Seventeenth-Century Thought." In *Psyche and Soma: Physicians and Metaphysicians on the Mind-Body Problem from Antiquity to Enlightenment*, edited by John P. Wright and Paul Potter, 197–215. Oxford: Clarendon, 2000.

Lokhorst, Gert-Jan. "Descartes and the Pineal Gland." In *Stanford Encyclopedia of Philosophy*, Winter 2017 ed. Stanford University, 1997–. https://plato.stanford.edu/archives/win2017/entries/pineal-gland/.

Luther, Martin. *Luther's Works*. Edited by Jaroslav Pelikan and Helmut T. Lehmann, 55 vols. St. Louis, MO: Concordia Publishing House; Philadelphia, PA: Fortress Press, 1955–1986.

Martin, Raymond, and John Barresi. *The Rise and Fall of Soul and Self: An Intellectual History of Personal Identity*. New York: Columbia University Press, 2006.

Mix, Lucas J. "Proper Activity, Preference, and the Meaning of Life." *Philosophy and Theory in Biology* 6 (2014). http://dx.doi.org/10.3998/ptb.6959004.0006.001.

Osler, Margaret J. *Divine Will and the Mechanical Philosophy*. New York: Cambridge University Press, 1994.

Osler, Margaret J. *Reconfiguring the World: Nature, God, and Human Understanding from the Middle Ages to Early Modern Europe*. Baltimore: Johns Hopkins University Press, 2010.

Ruse, Michael. *Darwin and Design: Does Evolution Have a Purpose?* Cambridge, MA: Harvard University Press, 2003.

Ruse, Michael. *Science and Spirituality: Making Room for Faith in the Age of Science.* New York: Cambridge University Press, 2010.

Voss, Stephen. "Descartes: Heart and Soul." In *Psyche and Soma: Physicians and Metaphysicians on the Mind-Body Problem from Antiquity to Enlightenment,* edited by John P. Wright and Paul Potter, 173–196. Oxford: Clarendon, 2000.

CHAPTER 13

Divided Hopes: Physics Versus Metaphysics

The mechanical philosophy created three serious problems for accounts of life. First, the loss of final causes made it difficult to explain "organisms." What is an organism, if it is not organized to some common end? Biological activities had previously been defined around their ends. Nutrition perpetuates individuals while reproduction perpetuates type. Sensation and locomotion feed both these basic ends and rational ends in humans. Rational life seeks truth.[1] Second, the loss of formal causes made it difficult to distinguish individuals and species from their surroundings. What distinguishes a body from a simple aggregate or lump of stuff? Third, mechanism cut the connections between vegetable, animal, and human life. The medieval syntheses had provided ways for mind to affect matter and vice versa. Souls had dealt with the proto-agency and unpredictability of living things as well as the full-blown agency and unpredictability attributed to minds.

The disjunction, now framed as the mind-body problem, raised additional challenges for monotheist concepts of spiritual life. The French philosopher Pierre Bayle (1647–1706) influenced many contemporaries through his widely read *Dictionnaire Historique et Critique*. He encapsulated the particular problem of animal souls. "Of all physical objects [of study], none is more abstruse or embarrassing than the bestial soul. The extreme views on this topic are either absurd or dangerous, the

---

[1] Recall that rational life is not subjectivity, but intellectual perception.

© The Author(s) 2018
L. J. Mix, *Life Concepts from Aristotle to Darwin*,
https://doi.org/10.1007/978-3-319-96047-0_13

mean that is sought between them is indefensible" ("Pereira," quoted in Lennon 2000, p. 211). Brute and human subjectivity appear very similar. If they are the same, then either brutes share in human transcendence or humans share in brute mechanism, lacking both interiority and true agency. The animal faculties resisted attempts to sequester them in either the material or the mental realm.

How to Study Life

Within the medieval syntheses, living things made sense as various levels of involvement in the life of the cosmos. One could speak of a continuous spectrum, running from base matter to divine order: the *scala naturae* or great chain of being. As the links proceed upward, they exhibit more and more the order of their spiritual anchor, God. All were pulled in both directions by the inherent hybridity of creation. More concretely, the Aristotelian synthesis accounted for life in terms of souls: the agreement of efficient, formal, and final causes. The study of all living things was properly called "psychology," the study of souls.

The Enlightenment ruptured this continuity. Spiritual and subjective life never fit easily into the Aristotelian continuum of souls. Nor did the use of final causes (after Alexander of Aphrodesias) allow for an easy understanding of mortal, natural souls. Medieval scholars had made the final cause into something otherworldly: intentions in the mind of God. Nominalism and mechanism brought these long-standing tensions to the surface.

A new science of life, in the sense of modern, mechanical biology, would not arise until the nineteenth century. Michel Foucault (1994, p. 207) stresses the late arrival of biology as a science. This should not be read to mean that it was never a unified endeavor. Psychology provided a unified theory and subject matter of life for Plato, Aristotle, and the medieval syntheses. Biology fell apart, however, in the Enlightenment because of the mechanical philosophy. Life, vegetable, and animal remained popular, understandable categories. One could still identify specialists who studied them and bodies of knowledge associated with them. And yet, no central theory or set of first principles tied them all together.

Philosophers and scientists attempted a dual epistemology. "Physiology" came to describe accounts of life based on physical, mechanical principles while "psychology" came to describe accounts of life based on subjective, mental principles. The mechanical philosophers placed biology at the edges of both. For most philosophers, life required a second, non-empirical epistemology to understand it. Life did not break the rules of nature, but it did require additional explanation. Bacon, Leibniz, and Kant each attempted to negotiate this boundary.

Bacon—Physics and Metaphysics

Francis Bacon (1561–1626), credited as the father of empirical reasoning, set forth his views on observation and induction in two works: *Of Proficience and Advancement of Learning Divine and Human* (2000) and *Novum Organum Scientarum* (2004). Both had a profound impact on the development of modern science, advocating for a Christian skepticism. Bacon argued that fallen humanity has lost its proper place in the universe. Science can restore our dominion over the created world (Dowe 2005, pp. 70–72). As an Anglican Christian and sometime Lord Chancellor of England, he was deeply entrenched in an English worldview that sought deeper understanding of God and morality through empirical study (Ruse 2003, pp. 34–37).

Bacon distinguished physics from metaphysics, placing material and efficient causes in the former, formal, and final causes in the latter (*Advancement*, 2.1). Physics focuses on the varied causes of particular events, while metaphysics focuses on fixed universal causes. Science involves working from particular to universal, from physics toward metaphysics, along the lines of the late Aristotelian synthesis. Bacon remained highly skeptical, however, of our ability to complete this movement. Final and formal causes provide the most desirable knowledge, but they are not practical. They cannot give us power over our environment.

Bacon called efficient causes the vehicles of final causes. The two never conflict (ibid., 2.2). One always acts through the other, making knowledge of efficient causes more useful. In the 1623 revision of *Advancement*, he compared final causes to virgins, who are consecrated to God. Bacon likely intended this to mean they were of utmost worth (as abstract knowledge) but never bear fruit (as practical knowledge).[2] Later biologists, notably Charles Darwin and T. H. Huxley, however,

---

[2] So William Whewell argued in his "Bridgewater Treatise" of 1833.

spoke of final causes as "barren virgins" in order to dismiss them as pointless.

Bacon proposed both spiritual and tangible substances. Following Aristotelian physics, he differentiated between the two, making the heavenly spheres entirely spiritual (Klein 2016). Below the moon, tangible objects are assembled from gross matter under the influence of more rarified, but still corporeal, air-like spirits. Organisms require more rarified, fire-like spirits, which distinguish vegetables, animals, and humans (*Novum Organum*, 2.40). These spirits fulfilled the same role as vegetable souls in Aristotle, empowering biological function. Bacon held that vegetable and animal souls (or spirits) are traduced, while human souls are specially created by God (Dales 1995).

Leibniz—Matter and Monads

Gottfried Wilhelm Leibniz (1646–1716) may have been the last of the Renaissance generalists, integrating history, philosophy, theology, natural science, and law. His work had a profound, if not always direct, influence on modern thought. He set forth his understanding of life in *Monadology* (1714).

Leibniz started with a dualism resembling Descartes'. A "kingdom of nature" operates as matter in motion, a mechanism designed by God. A second kingdom, the "kingdom of grace," encompasses souls and minds moved by final causes (as intention). Leibniz retained vegetable and animal souls, however. Souls are the primitive *entelechy* that makes matter into machines; they organize and drive it (Leibniz 1989, pp. 177, 222). Leibniz used the Aristotelian term *entelechy* in a new way, however. He described the soul as a substantial form, responsible for unity, nutrition, and reproduction, despite being epistemically dubious (pp. 82, 104, and 289). Troubled by mechanism, he wrote this.

> Therefore, I tried to fill this gap, and have at last shown that everything happens mechanically in nature, but that the principles of mechanism are metaphysical, and that the laws of motion and nature have been established, not with absolute necessity, but from the will of a wise cause, not from a pure exercise of will, but from the fitness of things. (p. 319)

Souls fill the gap between mechanical and spiritual, natural and gracious accounts of the world.

The two kingdoms move according to their own principles; they are each causally closed. And yet, by divine will, they coincide perfectly. He called souls *monads* or simple substances. Monads account for the coherence of all objects and are necessary to understand the properties of organisms. Further, each monad is composed of smaller monads, which have their own animal faculties of sensation and agency (p. 147).

Woven throughout this description was an anti-Platonic sensibility. Diversity expresses life better than participation. Thus, in place of the Neoplatonic cosmic soul, Leibniz spoke of the *identity of indiscernibles*, arguing that every particular is essentially unique. Nonetheless, he subscribed to a theory of reproduction in which living things only come from other living things of the same kind (p. 209). Within this apparent paradox, plants exemplify life and animals exemplify monads (Marder 2014, pp. 121–130). Plants grow indiscriminately with ambiguous boundaries, showing the profligacy of life. Animals have fixed bounds displayed in sensation (not-self moving self) and willed motion (self moving non-self). Flipping Ibn Sînâ's claim, Leibniz found plants clearly alive but only doubtfully ensouled.

Kant—Phenomena and Noumena

Immanuel Kant (1724–1804) integrated the a priori reasoning of Descartes with the empiricism of Bacon. A lifelong resident of Königsberg, he corresponded broadly and made significant contributions to philosophy and astronomy. His picture of the world and reasoning appears in three main "critiques." *The Critique of Pure Reason* defines metaphysics as reason independent of experience. It distinguishes between the *noumenal* world of "things in themselves" and the *phenomenal* world of appearances (appearing to noumenal observers). With Descartes, Kant believed that subjective experience was ontologically and etiologically prior to any physical reality. With Augustine, he thought the noumenal world existed beyond the range of certain knowledge. *The Critique of Practical Reason* defends the possibility of knowledge about God, freedom, and ethics. Some conclusions can be drawn in spite of the physical, metaphysical divide. *The Critique of Judgment* (*CJ*, 1952) looks at two problematic cases at the intersection: aesthetics and final causes.

Sections **63–68** in *CJ* deal with biological organization. Neither a priori reasoning nor observation reveals purposes within organisms; and yet we, as rational observers, must suppose that they exist (Chaouli 2017). Mechanical explanations are insufficient to understand organisms. "A plant is possible only in accordance with an idea. That exists only in the understanding and, for humans, in concepts" (Kant 2011, p. 329). That "idea" involves the ends of nutrition, growth, and reproduction along the lines of Aristotle's vegetable soul (*CJ*, 2: 63–64). Kant spoke specifically of *physical ends*, when something in nature is "both cause and effect of itself" echoing the Aristotelian identity of efficient and final causes. This applies both to nutrition, in continual production of self, and to reproduction, in copying according to genus.

Unity, organization, and identity fall out of this confluence of means and ends. More importantly for the history of biology, Kant suggested that final causes might be viewed diachronically as a nexus of efficient causes. Final and formal causes do not compete; they reflect the same phenomena viewed from different angles.

> What we require, therefore, in the case of a body which in its intrinsic nature and inner possibility has to be estimated as a physical end, is as follows. Its parts must in their collective unity reciprocally produce one another alike as to form and combination, and thus by their causality produce a whole, the conception of which, conversely,—in a being possessing the causality according to conceptions that is adequate for such a product—could in turn be the cause of the whole according to a principle, so that, consequently, the nexus of *efficient causes* might be no less estimated as an *operation brought about by final* causes. (*CJ*, 2: 64, italics in the original)

Kant even anticipated some aspects of inheritance, variation, and selection within the concept of physical ends (though he does not allow for change through time).

> But nature, on the contrary, organizes itself, and does so in each species of its organized products—following a single pattern, certainly, as to general features, but nevertheless admitting deviations calculated to secure self-preservation under particular circumstances. (*CJ*, 2: 56)

Physical ends cannot be justified through pure reason or observation. Instead they must be pragmatically assumed as a necessary precursor to reasoning about organisms.

DESIGN ARGUMENTS

The "argument from design" (to a designer) covers a broad class of Christian apologetics that uses biological complexity as a proof of God. Philosophers comment on the connection between biology and transcendent order or intent throughout Western history. And yet, it is not always clear which way the argument goes. Neoplatonic schemes take cosmic order and divine intent for granted and use them to explain biological complexity. Design arguments move from observed complexity to cosmic order. Nor do medieval arguments operate with the clean distinctions of the Enlightenment, between spiritual and physical, mind and matter.

The argument from design required two philosophical developments to make sense. First, it required the competition of efficient and final causes. Species either arise through a chain of mechanical efficient causes or through God's intervention as final cause. Organisms must be *more designed* than the rest of the cosmos. Second, the argument required that the intentionality of a designer, human or divine, be distinct from the physical universe. Human design must be supernatural (as artifice) in order to infer divine design (as miracle). It rests heavily on the ontological break between the physical and mental.

In this context, Cartesian dualism forced the design argument. If we take biological ends to be intuitively compelling but remove them from the physical world, they must go somewhere. For Enlightenment thinkers, that was the mind of God, a resting place prepared from the beginning of the Middle Ages to hold both knowledge and final causes. Gassendi, Bacon, Leibniz, and Kant all struggled to create a more nuanced philosophy with epistemic humility about final causes in a less dualistic universe. Nonetheless, the general migration of final causes to the mind of God during the seventeenth century resulted in a growing appeal to design which used biology as its chief witness.

Two writers became identified with the design argument as it developed in the Enlightenment. John Ray (1627–1705) spelled out the move from organisms to God in *The Wisdom of God, Manifested in the Works of Creation*. William Paley (1743–1805) more explicitly made the connection from clockwork to clockmaker in *Natural Theology, or Evidences of the Existence and Attributes of the Deity*. The latter had a strong influence on biological thought in the nineteenth century. It set up the problem that Charles Darwin sought to solve with evolution by natural selection.

IRRECONCILABLE DIFFERENCES

In the eighteenth and nineteenth centuries, many students of life argued for non-overlapping magisteria. Physics describes the extended world, metaphysics the world of proper ends, and never the twain shall meet. Popular summaries include the Is/Ought divide associated with Kant, the Why/How divide proposed by Charles Kingsley, and the Fact/Value divide defended by Stephen Jay Gould (1999, p. 4). Each of them glosses over the problematic areas of physiology and psychology that span the gap. Organisms have directionality, still spoken of as health and function.

As physiological and psychological accounts diverged, communities formed around each, leading to radically different terminology and epistemic frames. For much of the twentieth century, attempts to explain one in terms of the other were frowned upon, if not actively suppressed. It was not uncommon to see appeals for independence, pushing for reductive materialism and mechanism in physiology, reductive idealism in psychology. Thus, it can be hard to identify the true ontological dualists from reductionists of both sorts.

Setting humans to the side for the moment, four rough camps developed around questions of organization and proto-agency among plants and brute animals. Reductive idealists and reductive materialists accepted the Cartesian divide but tried to eliminate one realm or the other. "Vitalists," leaning Platonic, expanded the range of subjective or agential traits to animals, plants, and even minerals. They favored non-physical causes and set efficient and final causes in opposition. More Aristotelian thinkers emphasized immanent physiological processes that manifest psychological properties. They made final and efficient causes two aspects of the same thing.

Idealists

Before the Cartesian divide, the term idealism has little purchase. Most thinkers spoke of continua. Leibniz could only compare Plato's (relative) idealism to Epicurus' (relative) materialism, placing himself in between. Both medieval syntheses depended on an eternal *logos* etiologically prior to the physical universe, but this was not considered opposed to physical substance and causation. Christian Wolff (1679–1754) was the first to use the term "idealist" to classify philosophers. He identified Leibniz with idealism and Spinoza with materialism.

The new idealism had roots in the works of Leibniz and Nicolas Malebranche (1638–1715), a French Catholic priest and rationalist. Leibniz, as discussed above, saw subjective qualities in all things. All organized bodies are in some sense artificial; they are the product of agency in monads. Malebranche, like al-Ghazâlî, defended occasionalism, saying that God is the only agent in the universe, bringing about all things as their proximal efficient cause (e.g., Malebranche 2000, pp. 59, 111, 117; see also, Nadler 2000). God, he said, placed the plan for every generation within the first ancestors of every species. After that, they arose through the mechanical action of nature (Jolley and Scott 1997, pp. 174–179; 1980, p. 353). There is no physical necessity or law, only God continuously willing the laws of nature. Divine (final) causes cannot compete with natural (efficient) causes because all causes are divine.

Of non-human souls, Malebranche thought the only ambiguity was semantic (Malebranche 1980, pp. 492–496).[3] Plants and animals clearly have a corporeal motivator and complex behavior analogous to the spring and mechanism of a clock. And yet they have no internal agency. They display the intelligence of the Creator. They do not have spiritual or subjective life. Humans possess these through their soul. It receives impressions via animal spirits traveling through nerves to the brain (ibid., p. 49). It is also free to consent to or dissent from God's will and, thus, is necessary but not sufficient for true agency (Kremer 2000).

Malebranche approached epistemological idealism. He claimed we can have very little knowledge of the physical world; we can only reason from and through mental processes. George Berkeley (1685–1753), Irish bishop and philosopher, provided the clearest statement of ontological idealism. He argued against any material substance. Only subjective minds, capable of perception, exist. Only such minds can be the cause of action; matter, insofar as it is conceived, is only passive. He was happy calling such minds spirits and souls, but he limited them to God

---

[3] Later, Malebranche attributed a belief in subjective animal souls to Augustine, but claimed Augustine was following common prejudice and, if he had thought about it, would have realized he was wrong (p. 676). He added that the innocent suffering of animals would be inconsistent with divine justice; therefore, animals do not suffer.

and humans. With Malebranche he said that God (efficiently) causes all natural motion.

David Hume

Scottish philosopher and skeptic, David Hume (1711–1776) critiqued both rational and empirical epistemologies, inhibiting further idealism and beginning to unravel the design argument. Following ideas in both Ockham and Gassendi, he thought we may not reason to causation with certainty, or even probability. Shifting causation from an external event to a subjective judgment, he defended our ability to attribute efficient, but not final, causes.

In *Dialogues Concerning Natural Religion*, Hume accepted the appearance of function in organisms and the appearance of design in the universe but presented numerous effective critiques against the design argument (Sober 2008, pp. 126, 139–141; Ruse 2003, pp. 26–29). He found the analogy from organism to machine uncompelling. The two differ in both composition and operation. Further, the induction is weak; we have only one example of a designer (humans) and only one example of a cosmos to reason from. At best, organisms reveal a designer but nothing about that designer's properties.

Hume also critiqued the whole idea of faculties as unnecessary "causes" for observed activities. "These are only more learned and elaborate ways of confessing our ignorance" (Hume 1993, p. 65). He argued further that, if we must think in terms of activities, then nutrition and reproduction (vegetation and generation) explain natural order as well as reason (pp. 78–83). In this narrow sense only, the world can be thought of as an ensouled animal (with notably vegetable activities, p. 73).

The Iatromechanists

Physiology, meanwhile, drew on a developing ethos of experimentation and willingness to challenge ancient authority. The Flemish physician Andreas Vesalius (1514–1564) translated Galen's work from Greek into Latin and published a book on anatomy with detailed diagrams based on his own dissections. It marked a shift from Aristotelian and Stoic accounts to more structural and mechanical approaches.

The *iatromechanists*, a group of physicians favoring broadly mechanical approaches to medicine, followed the division of Descartes. Most felt that a mechanical physiology was sufficient to medicine. The soul likely existed but was irrelevant for treating disease. Among the most famous iatromechanists were Herman Boerhaave (1668–1738) and his student Albrecht von Haller (1708–1777). Both were substance dualists, largely agnostic about the relation between mind and matter, though leaning toward a Leibnizian pre-established harmony (Wright 2000). The group also included Giovanni Alfonso Borelli (1608–1679), Thomas Sydenham (1624–1689), Marcelo Malphigi (1628–1694), Friedrich Hoffmann (1660–1742), and William Cullen (1710–1790).

Authors known historically as physicians provided the most unequivocally mechanical accounts of bodies at this time (Leibniz 2016, p. xvi). Most supported an immaterial soul in humans for religious or ethical reasons, but a few vocal physicalists stood out. Julien Offray de La Mettrie (1709–1751) drew attention for his bluntly atheistic rejection of God and immaterial souls and his argument for a fully mechanical mind.[4] In spite of his ontological monism, however, he remained a dualist with regard to function. He attributed psychological functions to the nervous system and a "spring" or power source in the brain (Wright 2000). He attributed physiological functions to separate systems and other, lesser springs located around the body.

Reducible Life

Between the seventeenth and twentieth centuries trends in science and philosophy led to a tug of war between vitalism and mechanism. In biology, matter and mechanism were increasingly understood, but several problems remained, seemingly intractable. New "chemical" elements (e.g., carbon, nitrogen, oxygen) replaced the classical elements (i.e., earth, air, fire, and water) as the basic stuff of the physical world. Initially, chemists differentiated between biological or "organic" compounds and "inorganic" compounds, but the distinctions quickly disappeared. In 1828, Friedrich Wöhler stumbled on an abiological reaction to produce urea. This began a long series of experiments demonstrating the life-free production of complex organic compounds. The apex of this trend may be the Miller Urey experiment of 1952, which demonstrated the formation of sugars and other biological building blocks in what were

---

[4] *The Natural History of the Soul* and *Man a Machine* (1748).

then considered early Earth conditions. In modern chemistry "organic" refers to any chemistry with carbon–carbon bonds, including diamonds and graphite. The term "biochemistry" is now reserved for life, but it is unclear whether this signifies an objective distinction.

More complex material causes for life were proposed in the nineteenth century to describe the fundamental stuff of life. The most famous of these was "protoplasm." Definitions ranged from a vitalist fluid to a simple description of the liquid inside a cell. The latter definition won out and the term was eventually replaced by "cytoplasm."[5] The cytoplasm hosts a complex series of biological reactions called metabolism, which build up and break down biochemical molecules. Nutrition, stripped of agency and identity, became metabolism.

Irreducible Life

Despite great advances, the mechanical philosophy faced two major obstacles, one philosophical, one empirical. Philosophers were still concerned with proto-agency. Is human freedom, if it exists, ontologically distinct from the freedom of other organisms? Scientists were more concerned with questions of spontaneous generation. If life is simply mechanism, why can new life not be assembled from non-life?

The English physician William Harvey (1578–1657) explored circulation, previously explained in terms of souls or the inherent vitality of spirits.[6] He provided a way of thinking about nutrition—at least the distribution of resources within an organism—mechanically. He also provided the maxim that all life comes from eggs (*omne vivum ex ovo*). Francesco Redi (1626–1697) and Louis Pasteur (1822–1895) turned the maxim into experiments, demonstrating that spontaneous generation does not occur under normal conditions. Similar conclusions could

---

[5] And 'nucleoplasm' inside the nuclei of eukaryotes.

[6] Galen advocated three circulations of *pneuma*—nerves, arteries, and veins. See Chapter 6. Vesalius' student Realdo Colombo (1510–1559) unified arteries and veins in a single system passing the lungs. Blood, not *pneuma*, moves around the body. Similar ideas appeared earlier in the works of Ibn al-Nafis (1213–1288) and Michael Servetus (1511–1553), though neither theory received broad attention until the twentieth century. Harvey's key contribution was to note that food volume cannot account for blood volume, proving the existence of a constant medium, rather than simply the movement of nutriment from point to point (Osler 2010, pp. 140–142).

be drawn from the work of Matthias Jakob Schleiden (1804–1881) and Theodor Schwann (1810–1882) who promoted cell theory: cells are the basic unit of life; all organisms are composed of cells; and all cells come from other cells.

In the two millennia from Plato to Descartes, ambivalence about biology and physics was packaged in terms of souls. Those souls were sometimes construed as more ideal or spiritual and sometimes more material and physical, and yet they always marked the area of discomfort, where something more than physics but less than psychology seemed to be occurring. Descartes attempted to cut the Gordian knot and separate mind and matter with a mechanical philosophy.

Bacon, Leibniz, and Kant all tried to clarify the mechanical position. They bridged or denied the Cartesian ontological divide, but erected epistemological barriers with very similar results. Bacon distinguished practical, practicable physics from ideal but fruitless metaphysics. Leibniz invoked physical nature and gracious monads in miraculous harmony. Kant suggested phenomenal and noumenal worlds, one apparent to the senses, the other only experienced internally. Together, they lay the foundations for the divided epistemology of natural science and the humanities.

Within the new paradigm of natural science, mechanical biology proved immensely successful. It revealed material and proximate efficient causes (or at least correlates) for nutrition, reproduction, sensation, and locomotion. Reason, subjectivity, and spirituality, meanwhile, remained out of bounds. At the level of chemical elements, life looked like non-life, but the problem of organization remained. How do the elements assemble into an organism? Life comes only from other life. Philosophers and scientists considered a number of possibilities. Those that ventured too far from the mechanical norm attracted the name of "vitalism," discussed in the next chapter.

REFERENCES

Bacon, Francis. *The Advancement of Learning*, vol. 4. Edited by Michael Kiernan. The Oxford Francis Bacon. Oxford: Oxford University Press, 2000.

Bacon, Francis. *The Instauratio Magna Part II: Novum Organum and Associated Texts*. Edited by Graham Rees and Maria Wakely. Oxford Francis Bacon 11. Oxford: Oxford University Press, 2004.

Chaouli, Michel. *Thinking with Kant's "Critique of Judgment."* Cambridge, MA: Harvard University Press, 2017.

Dales, Richard C. *The Problem of the Rational Soul in the Thirteenth Century.* New York: E. J. Brill, 1995.

Dowe, Phil. *Galileo, Darwin, and Hawking: The Interplay of Science, Reason, and Religion.* Grand Rapids, MI: Eerdmans, 2005.

Foucault, Michel. *The Order of Things: An Archaeology of the Human Sciences.* New York: Vintage, 1994.

Gould, Stephen Jay. *Rocks of Ages: Science and Religion in the Fullness of Life.* New York: Ballantine Publishing Group, 1999.

Hume, David. *Dialogues and Natural History of the World.* New York: Oxford University Press, 1993.

Jolley, Nicholas, and David Scott, eds. Malebranche: Dialogues on Metaphysics and on Religion. New York: Cambridge University Press, 1997.

Kant, Immanuel. *The Critique of Judgement.* Translated by James Creed Meredith. New York: Oxford University Press, 1952 [*CJ*].

Kant, Immanuel. *"Observations on the Feeling of the Beautiful and Sublime" and Other Writings.* Edited by Patrick Frierson and Paul Guyer. Cambridge Texts in the History of Philosophy. New York: Cambridge University Press, 2011.

Klein, Jürgen. "Francis Bacon." In *Stanford Encyclopedia of Philosophy*, Winter 2016 ed. Stanford University, 1997–. https://plato.stanford.edu/archives/win2016/entries/francis-bacon/.

Kremer, Elmar J. "Malebranche on Human Freedom." In *Cambridge Companion to Malebranche*, edited by Steven Nadler, 190–219. New York: Cambridge University Press, 2000.

Leibniz, Gottfried Wilhelm. *Philosophical Essays.* Translated by Roger Ariew and Daniel Garber. Indianapolis: Hackett Publishing Company, 1989.

Leibniz, Gottfried Wilhelm. *The Leibniz-Stahl Controversy.* Translated by François Duchesneau and Justin E.H. Smith. The Yale Leibniz. New Haven: Yale University Press, 2016.

Lennon, Thomas M. "Bayle and Late Seventeenth-Century Thought." In *Psyche and Soma: Physicians and Metaphysicians on the Mind-Body Problem from Antiquity to Enlightenment*, edited by John P. Wright and Paul Potter, 197–215. Oxford: Clarendon, 2000.

Malebranche, Nicolas. *The Search After Truth* and *Elucidations of the Search After Truth* and *Philosophical Commentary.* Translated by Thomas M. Lennon and Paul J. Olscamp. Columbus: Ohio State University Press, 1980.

Malebranche, Nicolas. *Dialogues on Metaphysics and on Religion.* Edited by Nicholas Jolley and David Scott. Cambridge Texts in the History of Philosophy. New York: Cambridge University Press, 2000.

Marder, Michael. *The Philosopher's Plant: An Intellectual Herbarium.* New York: Columbia University Press, 2014.

Nadler, Steven. "Malebranche on Causation." In *Cambridge Companion to Malebranche*, edited by Steven Nadler, 112–138. New York: Cambridge University Press, 2000.

Osler, Margaret J. *Reconfiguring the World: Nature, God, and Human Understanding from the Middle Ages to Early Modern Europe*. Baltimore: Johns Hopkins University Press, 2010.

Ruse, Michael. *Darwin and Design: Does Evolution Have a Purpose?* Cambridge, MA: Harvard University Press, 2003.

Sober, Elliott. *Evidence and Evolution: The Logic Behind the Science*. New York: Cambridge University Press, 2008.

Wright, John P. "Substance Versus Function Dualism in Eighteenth Century Medicine." In *Psyche and Soma: Physicians and Metaphysicians on the Mind-Body Problem from Antiquity to Enlightenment*, edited by John P. Wright and Paul Potter, 237–254. Oxford: Clarendon, 2000.

CHAPTER 14

Ghosts in the Machine: Vitalism

Questions of organismal unity and purpose remained problematic under the mechanical paradigm. Alternative paradigms emerged. Though wildly different, they all attracted the label of vitalism. One group, identifying themselves as mechanists, accepted the ejection of formal and final causes from biology but sought to identify life-specific material and efficient causes. A second group, identifying themselves as vitalists, proposed vital spirits that were more than matter in motion but less than souls. A third group, identifying with Idealism and Romanticism, pushed for more subjectivity and spirituality within the nonhuman world.

The term vitalism can be difficult to pin down. Roselyne Rey (2000, p. 255) defines it as "a view on the relationship of body and soul whose main feature may be seen as a rejection of any form of simplistic dualism." Canguilhem (2008, p. 60) claims, "the term vitalism is appropriate for any biology careful to maintain its independence from the annexationist ambitions of the sciences of matter." In the twentieth century, vitalism became an insult aimed at any biological theory deemed outside the bounds of proper empiricism, particularly when it invoked teleology or nonphysical causation. Because the boundaries of mechanical reasoning were (and are) under debate, the term can be problematic. It includes some theories disallowed a priori and some disproven a posteriori, without differentiating between the two. In any case, this chapter looks at a variety of broadly Platonic life-concepts between Descartes and Darwin.

© The Author(s) 2018
L. J. Mix, *Life Concepts from Aristotle to Darwin*,
https://doi.org/10.1007/978-3-319-96047-0_14

THE ANIMISTS

A few biologists still invoked nonhuman souls. The German chemist and physician, Georg Ernst Stahl (1659–1734) was the most famous *animist* in this period. He believed that mechanical processes cannot respond quickly and specifically enough to maintain organized bodies. Organisms are materially and chemically unstable, requiring an immaterial formal, efficient, and final cause to hold them together. Stahl's position developed in opposition to the mechanism of Boerhaave and Hoffmann. His contentious and extensive correspondence with Leibniz provides a broad exploration of his biological philosophy (Leibniz 2016).

While Leibniz had associated life with sensation and appetite. Stahl associated it with a capacity to resist decay (p. lxi).[1] He argued for vegetable souls using a mechanical clock metaphor to contrast machines and organisms. The clock has been designed by a human. It acts as an organism by fulfilling its purpose: to tell time accurately. As an extension of its designer, it is both organic and instrumental. When the clock breaks, it continues to follow the laws of physical necessity, but it is no longer instrumental. It becomes simply mechanical. The human provides the clock with formal, efficient, and final causes, making it organic, but only when it runs properly. An organism occurs when the formal, efficient, and final causes exist within the clock: when it defines, creates, directs, and copies itself. Stahl attempted to naturalize formal and final causes by moving them out of the mind of God and back into a vegetable soul.

Stahl followed the idealist picture of agency in God and souls. God moves matter through the course of natural laws, while souls motivate humans within the larger system (p. l). He extended souls to all organisms, however, returning to the Aristotelian definition. A soul is "a principle energetically endowed with the faculty of moving" (p. liii). Nutrition provides the most fundamental organismic motion. Stahl described it as the process of appropriating particles to proportionally and relationally form organs. Development (growth) and reproduction both proceeded from nutrition, as methods of ordering and purposing matter. He referred to the immaterial cause of nutrition as both *anima* and *natura* in line with the Greek plan of *psyche* as the *physis* of a living being.

---

[1] For original sources, see Stahl's *New Medical Theory* (1708), in particular, the introductory essay "Disquisition on the difference between mechanism and organism."

Stahl still had to account for ideas about subjective life, so central to Leibniz and other contemporaries. For this, he distinguished a *logos* or *ratio* (providing vegetable order and activity in each organism) from a *logismos* or *rationatio*, which provides for critical thinking. The *logos* has no imagination or reason, but still regulates an organic body. The *logismos* possesses both and provides conscious rational regulation. Animal faculties remained, problematically, in between.

THE IATROCHEMISTS

Some physicians rejected both mechanical and animist arguments. The *chemical philosophy* developed in the Renaissance around the ideas of Paracelsus (Theophrastus von Hohenheim) a Swiss physician and alchemist living 1493–1541 (Osler 2010, pp. 118–124; Foucault 1994). Predating Descartes and Gassendi, he advocated for a natural philosophy with strong roots in both Stoic and Platonic philosophy. The chemical philosophy maintained the cosmic interconnections favored in Neoplatonism and, thus, allowed for action at a distance in ways unimaginable to the mechanists.

Jan Baptiste van Helmont (1580–1644), a Flemish physiologist and chemist, prominently advocated for the chemical philosophy among physicians. Living at the critical point when alchemy shifted to modern chemistry, he proposed an empirical approach to physiology but invoked both sensitive souls and *archei*. The *archei* were lower vital-principles that manage the chemistry of the body below the sensitive level.

THE (IATRO)VITALISTS

A group of French physicians calling themselves *vitalists* disagreed with previous life-concepts on methodological grounds (Rey 2000). They attempted to unify physiological and psychological activities. Mental states clearly affect the body; the body clearly constrains the mind. They rejected the machine metaphor, thinking it required external motivation. They also rejected animism. J.J. Méneuret de Chambaud (1733–1815) dismissed any appeal to a soul: vegetable, animal, or rational.

> To refer to the soul for an explanation of life and as a place to discover the laws of the animal economy is to cut the knot rather than to untie it, to

beg the question, and to enshroud it in the same obscurity this spiritual being (the soul) itself possesses for us. (quoted in Rey 2000, p. 257)

Méneuret cited both Gassendi and van Helmont as unacceptable examples of begging the question.

The vitalists invoked a third thing, neither body nor soul. Henri Fouquet (1727–1806) drew on the history of the sensitive soul and spoke of *sensibility* (French *sensibilité*) as a faculty intermediate between body and soul as then conceived. He cited van Helmont but indicated that sensibility includes all the vital functions, both conscious and unconscious. In general, vitalists returned to the classical debate about the unity and plurality of life-concepts. They leaned toward substance monism with functional dualism as in the Aristotelian synthesis. One type of thing accounts for a variety of processes. Méneuret spoke of "vital" or "natural" functions as distinct from "animal" functions. Marie Bichat (1771–1802) invoked "organic" versus "animal" sensibility. Paul Joseph Barthez (1734–1806) remained agnostic about the vital principle. He rejected the idea that it is a substance but believed that some kind of "energy" must explain organismality, unity, and self-regulation.

Vital Forces and Fluids

The question of vital forces deserves special attention. Newtonian theories expanded the mechanical philosophy beyond the mechanical metaphor, introducing the action of forces. These forces allowed for some kind of action at a distance. Etiological commitments to matter in motion remained strong, as did the ejection of formal and final causes.

Theories of gravity, electricity, and magnetism matured in this period—and radioactivity shortly after. Many natural scientists asked whether life might be defined by one of these forces. Or, perhaps it represented an additional force, not yet characterized. The German physician Franz Mesmer (1734–1815) proposed *animal magnetism* or *mesmerism*, a force similar to magnetism that flows within the human body and effects health. Luigi Galvani (1737–1798) suggested *animal electricity*, later called *galvanism*. Lorenz Oken (1779–1851) expanded on galvanism, drawing heavily on Aristotle. His work, *Elements of Physiophilosophy* defines life along classical lines. "Motion is therefore the soul, whereby the Organic is elevated above the Inorganic" (Oken 1847, sec. 992). Oken also set forth three kingdoms—mineral, vegetable, and animal—defined on the basis of composition (secs. 463–483). Using the classical four elements, he said that minerals contain only two (earth and one other). Vegetables contain three: earth, water, and air. Each element corresponds to galvanism and a classical faculty: nutrition, digestion, and respiration respectively. Animals contain all four elements and, thus, best represent an organic cosmos. Oken argued that the vegetable faculties are involuntary, but animal faculties involve both sensation and will (sec. 1012).

Luis Campos traces a similar, if smaller trend in his book *Radium and the Secret of Life*. After the discovery of radium by Pierre and Marie Curie in 1898, several researchers explored the possibility of a life-force analogous to radiation. Remnants of vital force thinking can also be seen in the twentieth century work of Henri Bergson (1859–1941, *élan vital*) and Hans Driesch (1867–1941, *entelechy*).

German Idealism

In Germany, philosophy, medicine, and the natural sciences intertwined and cross-pollinated, making it difficult to separate various trends. Central to our purposes will be Romantic poetry, nature-philosophy (*Naturphilosophie*), and natural science (*Naturwissenschaft*). All three integrate physiology and psychology by extending subjectivity, intention, and agency throughout the nonhuman world. Proponents saw themselves as followers of Leibniz and Kant but emphasized the subjective character of life while downplaying both mechanism and skepticism.

In broad strokes, the Germans represented a return to the fluidity of Heraclitus. Berkeley had denied that minds are material, but the German Idealists argued we should not even ask the question. Instead, we should speak of the continuity of minds and bodies with their environment. They rejected what they saw as idealization, isolation, and reification of souls in Kant and others (Miller 2002, p. 97). They associated plants with nonindividuality, passivity, and the feminine. Brute animals have a soul (*seele*). Plants may not, but they still participate in a spirit or mind (*geist*) that defines both life and subjectivity.

Johann Wolfgang von Goethe (1749–1832), a poet and philosopher with a strong interest in botany, was among the first to write in this vein. He invoked an *Urplant*, the shared plan, present all plants, and a similar type for animals (Ruse 2003, pp. 66–67). He also argued that life must be viewed as a process in time. Goethe rejected both the purely external motivation of the mechanists and the purely internal motivation of the animists (ibid., p. 63).

The central German Idealists were Johann Fichte (1762–1814), Friedrich Schelling (1775–1854), and Georg Hegel (1770–1831). They compared Nature to an unconscious organism, with conscious Spirit growing within it (Martin and Barresi 2006, p. 187). Light, gravity, magnetism, electricity, and biological organization arise as the precursors or first movements of that Spirit (Ruse 2010, pp. 151–152). Schelling invoked an organic holism in which parts bring about the whole, but the whole also brings about the parts. Echoing design arguments, he thought that inherent purpose and self-organization can be observed in individual organisms and in Nature at large.

This association of animal and natural teleology developed into an idea of progressive evolution, influencing Oken as well as Richard Owen (1804–1892), Ernst Haeckel (1834–1919), and others (Ruse 2010, pp. 150–155).[2] German Idealism kept the cosmic dynamism of Neoplatonism but changed it from an eternal flow to a historical process. This suggested that species might change through time; they were no longer eternal kinds. It also suggested a global intentional teleology; the Spirit moves life in some foreseen direction. Michael Ruse (2010, p. 153) argues that this teleology was more Aristotelian than Platonic; it invoked an internal purpose rather than an external one. In the treatment of individual animals, this was strictly true; but, by speaking of organic Nature, the nature-philosophers also brought back Plato's cosmic soul. They saw a drive (internal to the cosmos) directing (external to organisms) the development of species. Examples of such a drive within evolution include the *orthogenesis* of Wilhelm Haacke (1855–1912), the *noogenesis* of Pierre Teilhard de Chardin (1881–1955). This is not to say that the move from Schelling to Chardin was direct or inevitable, only to mark a point of connection in the long history of life-concepts.

The German Idealists attempted to reintegrate final causes and the physical world and, in so doing, brought back Platonic transcendence. They perceived global trends, pushing organisms from simplicity to complexity. Several described parallel trends in animals, running from animalcules to higher vertebrates, and in plants, running from lichen to trees.

---

[2] According to Wallace (1910, p. 333), Haeckel used a soul-particle along the lines of Lucretius.

Georg Hegel

In *Philosophy of Nature*, Hegel resurrected the Platonic hierarchy as a historical narrative or movement of Spirit (Miller 2002, pp. 124–126). He did not feel bound to the epistemology of science, however, and focused on the process of human meaning-making. Section 3 details three organic realms within nature: the mineral, plant, and animal. Minerals are "organic" (meaning structured) but inanimate. They have no subjectivity. Plants have neither interiority nor subjectivity but possess a minimal agency because nutrition allows them to negate "the other," turning it into self. Marder (2014, pp. 155–160) describes their liminality as "passive masters." "Akin to the fermenting grapes, suspended between nature and culture, the plant is a dialectical point of transition" (p. 159).

For Hegel, plants exist at the threshold of life in three ways: subjectivity, individuation, and struggle (Hegel 2004, 3:2.266–272). As a subject, plants have no consciousness, soul, or interiority. Passive, they make no I-other distinction, despite being constantly in the process of turning other into self. As individuals they constantly diverge, existing more as aggregates than unified wholes. They proliferate as much as they grow. Because they cannot strive or oppose, they exist as a stepping stone to a greater expression of spirit. "The plant is a subordinate organism whose destiny is to sacrifice itself to the higher organism and to be consumed by it" (ibid., p. 350). Drawing on antique and medieval dichotomies, Hegel emphasized a pairing between the active, oppositional, animal masculine and the passive, non-oppositional, vegetal feminine (Miller 2002; Marder 2014, pp. 153–170).

Animals, for Hegel, were clearly alive and ensouled. Sensation and willed motion demonstrate and require differentiation from the environment. He thought that the development of true self-consciousness requires this struggle (Martin and Barresi 2006, p. 188). It proved integral to Hegel's dialectical reasoning. In nature, plants are thesis, animals antithesis, and humans synthesis in one great movement of Spirit.

Friedrich Nietzsche

German philosopher and poet Friedrich Nietzsche (1844–1900) went a step beyond Hegel. Taking the rejection of formal and final causes seriously, he emphasized the constant flow of Heraclitus. Nietzsche denied all individuality and teleology (Miller 2002, p. 151).[3] Everything was in a constant state of flux and metamorphosis. "There is no form in nature, for there is no inner and no outer" (quoted on p. 152). Nietzsche opposed vitalism because he could not see a clear distinction between life and nonlife. He rejected Kant's noumena and physical ends as unjustifiable. He argued that unity, consciousness, and subjectivity are merely illusions. We have fooled ourselves with our language into thinking of these as causes, much less substances. Nor are atoms and elements any more substantial.

Drawing on Goethe, Nietzsche used plants as a sign of transgressive, unbounded life (pp. 153–155). He spoke of metamorphosis as a constant transitioning of everything into everything else. Numerous twentieth-century philosophers followed his lead and used plant metaphors to critique traditional concepts of identity, individuality, and unity. Notable examples include the *dissemination* of Jacques Derrida (1930–2004), the *rhizomes* of Gilles Deleuze (1925–1995) and Félix Guattari (1930–1992), and the *efflorescence* of Luce Irigaray (1930–).

THE TWO NATURES

A constant thread in the mechanism/vitalism debates was argument over the extent of "nature." John Stuart Mill (1806–1873), a British thinker known for his political philosophy, stated the problem concisely. We get in trouble when we slip unknowingly from one meaning of "nature" to another.

> It thus appears that we must recognize at least two principal meanings in the word Nature. In one sense, it means all the powers existing in either the outer or the inner world and everything which takes place by means of those powers. In another sense, it means, not everything which happens, but only what takes place without the agency, or without the voluntary and intentional agency, of man. (Mill 1885, p. 8, "Nature")

---

[3] Miller (2002, p. 172) quotes Nietzsche specifically on the definition of life, plants, and individuation. "In order to understand what 'life' is, what kind of striving and tension life is, the formula must apply equally well to trees and plants as to animals. 'What does a plant strive after'—but here we have already invented a false unity that does not exist: the fact of a millionfold growth with its own and quasi-its-own initiatives is concealed and denied if we posit a crude entity 'plant.'"

We might take nature in the all-encompassing sense as everything that happens. This would be in line with the physis of Aristotle and the *natura* of Aquinas. In this sense, souls, agency, and biological activity must, by definition, be natural. If we take this path, then we must describe physics in a way that includes accounts of what we label nutrition, reproduction, sensation, will, etc. Alternatively, we might take nature in a restricted sense that excludes these things as supernatural or artificial. This would be in line with the Platonic star-soul, the dual creation accounts of medieval Christianity, and the *res cogitans* of Descartes. Our challenge, then, will be to understand when, where, and how transcendent events interact with nature.

In neither case can we "naturalize the soul." If nature is all inclusive, it includes whatever animation and agency humans possess. If nature is everything physical but humanity, then we have a priori committed to something un-natural about humanity. Confusion only arises when we define nature and agency exclusively, then look for agency in nature. The same challenge applies to nutrition, reproduction, sensation, and will. We must either define nature to include such causal nexuses, distinct from simple matter in motion, or not.

The ambiguity of vegetable and animal life-concepts hangs on this confusion. In the Aristotelian case, human nature must arise within a more general nature, most likely a vegetable and animal nature with increasing levels of freedom. In the Platonic case, humans possess unique agency because of their transcendence, but there is no a priori reason this would not be true of all animals and plants (with decreasing levels of freedom). The problem only becomes intractable when we insist on the division.

This was precisely the tack taken by Descartes. The "naturalization of the soul," or more precisely the naturalization of physiology and psychology in the eighteenth and nineteenth centuries reflected an attempt to undo the Cartesian divide, but it would be anachronistic to project this divide backward without careful analysis. Most of the influential life-concepts attempted to tackle this very problem. Indeed, the soul might be described as a millennial attempt to bridge materiality and the dynamism of living things. The "vitalists" were struggling to work out a new synthesis.

EMERGENCE

Mill attempted to find a middle way between vitalism and mechanism with a new approach called *emergentism*. The British emergentists argued that there is only one kind of substance, the substance of particles and physics, but that some aggregates have additional properties that cannot be reduced to the primary qualities of the mechanical philosophers (O'Conner and Wong 2015).

George Henry Lewes proposed the term 'emergence' in *Problems of Life and Mind* (1874, 1875). Physiology, he said, deals with the context of living systems, psychology with the consequences. He worried that "matter, force, cause, life, mind, object, and subject" had been excluded as nonempirical metaphysics (1:57). They can be discovered, however, as relational properties discerned a posteriori. One must do physics before one can do metaphysics but, having observed phenomena, it is possible to speak of laws behind the observed regularities.

With regard to biological activity, Lewes saw physiological and psychological accounts as different conceptions of the same thing (2:418).[4] "An agent can be that agent only in its agency, a stimulus is such only when it stimulates... The object isolated in abstraction is a *possible* agent but is not *really* an agent at all apart from the co-operant organ" (2:419, italics in the original). Thus, organisms are real and distinct, just as shadows and molecules are real and distinct. They have properties that can only be understood as relationships between parts. Like molecules, they have properties, even causal power, which could not be predicted a priori. In this way, he thought that individual organisms are real, but collections of organisms (species, genera, etc.) are not (2:44).

Mill echoed Lewes, arguing that nutrition is as much a "law of nature" as gravitation, though it occurs locally in organisms and not universally (Mill 1885, p. 6). When we place two weights together, their total mass is equal to the sum of their individual masses. But, when we join hydrogen and oxygen, they develop a dipole moment, a separation of positive and negative charge, not previously apparent. Both "laws" (gravitation and dipole formation) are universal, but the latter only manifests in specific, relational circumstances.[5]

---

[4] Lewes was thinking specifically of the relation between neural excitation and sensation.

[5] The dipole moment and many properties of water are now a priori predictable based on the relative attraction of electrons to the nuclei of both atoms. Broad suggested a more remote example. The scent of ammonia would not be predictable without an

Lewes, Mill, and C.D. Broad (1887–1925) expressed what has been called *supervenience*. They believed that some aggregates have novel causal powers or *top-down causation*. In their view, entities can be real at different levels. Many modern emergentists make the weaker claim that some phenomena can be explained at a lower level in principle but not in practice. The lowest level remains exclusively substantial and causal while higher levels become useful descriptions.

In the eighteenth and nineteenth centuries, many new Platonic syntheses were proposed. They have largely been rejected and labeled "vitalist" but many were important to the development of later biology. Philosophers and physiologists sought to supplement a mechanical foundation with life-specific substances and forces. *Archei, sensibilité, mesmerism, galvanism, élan vital, entelechy,* and *geist* were all proposed to fill the gap left by souls.

They fell into the same trap as the subsistent souls of the Middle Ages: circularity. If we start by defining nature (or physics or science) in a way that excludes life, we cannot leverage that definition to discover new properties at the boundary. Such Platonic approaches may appeal to the intuition, but they will never meet Bacon's requirement for science: pragmatic predictions. To the extent that they succeed as a philosophy of biology, they must diverge from natural science under the mechanical paradigm. This need not be a bad thing, but it bears close attention, lest we return to the confusion mentioned by Mill. We must be clear what we mean by nature.

The vitalists tried to reintegrate subjectivity, sensation, will, and reason into a unified picture of the world. The mechanical philosophers, meanwhile, cordoned off precisely these issues so that the remaining matter in motion might be tractable to explanation. Breaking those boundaries risked losing both the interesting characteristics of life—its unique causal powers—and the predictive power of "natural" science.

Luckily, a new Aristotelian approach to the physics/life gap also arose at this time: evolution. Though it did not fully bridge the divide, it brought biology one step closer to the mechanical finish line. Jumping

---

understanding of the sensory apparatus, both mechanical and subjective, of the observer. It is not a property of the ammonia, but the ammonia/observer system. Recent discussion of emergence explores the extent to which this is meaningfully distinct from reductionism. See O'Conner and Wong (2015, 3.3).

off from the causal nexus of Kant and the historicized development of nature-philosophy, proponents of evolution would build upward from material and efficient causes to explain organismality, function, and biological activity.

REFERENCES

Canguilhem, Georges. *Knowledge of Life*. New York: Fordham University Press, 2008.

Foucault, Michel. *The Order of Things: An Archaeology of the Human Sciences*. New York: Vintage, 1994.

Hegel, G.W.F. *Hegel's Philosophy of Nature*. Translated by A.V. Miller. Oxford, UK: Clarendon, 2004.

Leibniz, Gottfried Wilhelm. *The Leibniz-stahl Controversy*. Translated by François Duchesneau and Justin E.H. Smith. The Yale Leibniz. New Haven: Yale University Press, 2016.

Lewes, George Henry. *The Problems of Life and Mind*, 2 vols. Boston: Osgood and Company, 1874, (1875).

Marder, Michael. *The Philosopher's Plant: An Intellectual Herbarium*. New York: Columbia University Press, 2014.

Martin, Raymond, and John Barresi. *The Rise and Fall of Soul and Self: An Intellectual History of Personal Identity*. New York: Columbia University Press, 2006.

Mill, John Stuart. *Nature, the Utility of Religion, and Theism*, 3rd ed. London: Longman, Green, 1885.

Miller, Elaine P. *The Vegetative Soul: From Philosophy of Nature to Subjectivity in the Feminine*. Albany, NY: State University of New York Press, 2002.

O'Connor, Timothy, and Hong Yu Wong. "Emergent Properties." In *Stanford Encyclopedia of Philosophy*, Summer 2015 ed. Stanford University, 1997–. https://plato.stanford.edu/archives/sum2015/entries/properties-emergent/.

Oken, Lorenz. *Elements of Physiophilosophy*. Translated by Alfred Tulk. London: C. and J. Adlard, 1847.

Osler, Margaret J. *Reconfiguring the World: Nature, God, and Human Understanding from the Middle Ages to Early Modern Europe*. Baltimore: Johns Hopkins University Press, 2010.

Rey, Roselyne. "Psyche, Soma, and the Vitalist Philosophy of Medicine." In *Psyche and Soma: Physicians and Metaphysicians on the Mind-Body Problem from Antiquity to Enlightenment*, edited by John P. Wright and Paul Potter, 255–265. Oxford: Clarendon, 2000.

Ruse, Michael. *Darwin and Design: Does Evolution Have a Purpose?* Cambridge, MA: Harvard University Press, 2003.

Ruse, Michael. *Science and Spirituality: Making Room for Faith in the Age of Science.* New York: Cambridge University Press, 2010.

Wallace, Alfred Russel. *The World of Life: A Manifestation of Creative Power, Directive Mind and Ultimate Purpose.* London: Chapman and Hall, 1910.

CHAPTER 15

The Same and Different: Early Theories of Evolution

Evolution began to fill the gap left by vegetable souls. It provided an alternative explanation for biological organization and a new first principle for biological accounts: adaptation. A broad definition of evolution—as change through time—follows naturally from the historical thinking in German Idealism.[1] Idealism and increasing knowledge of developmental biology contributed to a new, progressive evolutionary mindset, one that saw change through time as a movement from simpler to more complex and less valuable to more valuable. Progressive evolution aligned well with both the Platonic hierarchy and divine providence. And yet it failed on precisely those grounds. The values and goals of the universe remain obscure. Whether nonexistent or simply unknowable, they cannot contribute to an empirical biology. The broad definition of evolution caught the popular imagination—and the scientific imagination—but lacked the concrete explanations and predictions necessary to anchor biology as an empirical pursuit along the lines of Bacon, Leibniz, and Kant.

---

[1]Following the history of the words more strictly, "evolution" arose in the early seventeenth century from the Latin word for "unrolling" as of a scroll. Originally it referred to a change in military formation. It quickly took on a connotation of growth and development. Bonnet used it in 1762 to refer to *preformationism*. Darwin avoided the term, but it became popular in reference to his non-progressive theory by the end of the nineteenth century.

© The Author(s) 2018
L. J. Mix, *Life Concepts from Aristotle to Darwin*,
https://doi.org/10.1007/978-3-319-96047-0_15

The new life-concept rested on three crucial changes to thinking about organisms. First, the hierarchy of life became a historical process, situated in time and space. Second, divine agency and intention gave way to interactions between the environment and organisms. Third, the vegetable, animal, and human categories were reconnected in a way that emphasized their common vegetable activities: nutrition and reproduction.

CLEARING THE PATH

The writings of John Locke (1632–1704) and Auguste Comte (1798–1857) demonstrate a major shift in epistemology, clearing the way for Darwin's evolution. Both gave up appeals to a transcendent intellect (an infallible rational faculty) in favor of appeals to the senses (fallible animal faculties). A British philosopher, Locke (1824) addressed the limits of empirical reasoning, rejecting intellect and rational-life as a category. "Reason" became a term for subjectively processing sense data. Those who continued in this vein emphasized human life as subjective and spiritual, but not possessed of a unique form of perception.

Locke returned to the Ancient Greek question of what persists through change.[2] He distinguished between three basic substances: bodies, minds ("finite intelligences"), and God. With Aristotle, he defined the first two in terms of continuous processes. Inanimate "bodies," he says, do not persist. They cease to be the same when parts are added or lost. Add or remove a grain of salt from a block and you have a new body. Plant bodies are different; they absorb and excrete matter continuously. Vegetable identity comes from organization and nourishment persisting through changing matter. Animal identity requires, additionally, internal ordering and motivation.

Locke introduced an important distinction. "Human" identifies the brute animal individuality of a body, a continuity of order and nutrition. "Person" refers to something else, the continuity of consciousness. In this way, he saved the human body for physics and physiology, while preserving subjective personhood for immaterial souls. A human is only, and specifically, a kind of animal. A person is something else, a metaphysical

---

[2] *Essay Concerning Human Understanding*, 2.27; see also *Elements of Natural Philosophy*, Chap. 9 for discussion of plants and nutrition.

thing with consciousness and agency and, therefore, culpability (*Essay Concerning Human Understanding*, 2:27.26; see also 2:21.5–6).

Nutrition makes sense of all living bodies, from the lowest up to humans. Locke saw a continuous spectrum and even suggested that reproduction does not preserve species (3:6.12, 23). Species, in biology and other natural sciences, refer only to the names we attribute to the world. They contain no information other than this, as we can see from the production of hybrids (e.g., mules). Further, not even plants and animals are essentially distinct, but that we name them so.

The French philosopher Auguste Comte defended empirical positivism, the idea that things may be known with certainty based on the observation. This represented a shift away from the skeptical empiricism of Bacon, Hume, and Locke. His progressive theory of knowledge described human societies as moving through stages. In the theological stage, we look for formal and final causes and attribute events to personal agents. In the metaphysical stage, we look for abstract (impersonal) entities that act as efficient causes. In the positive stage, all causes are given up in favor of universal laws. This was demonstrated in biology by *animism* (in the sense of personal spirits, such as dryads in streams), vegetable souls, and finally a physics of the interaction between organisms and their environment. Comte (2001, book 5) saw recent history as a tug-of-war between physical and metaphysical explanations in biology. He rejected vitalism and iatrochemical approaches as metaphysical, but also rejected mechanism as too reductionist. Life is a process, and nutrition is its most fundamental and generic expression.

Neither Locke nor Comte, nor a broad range of philosophers in between, accepted the full mechanization of plants. They did accept Bacon's judgment of metaphysics. Material and efficient causes should be sufficient for biology, but only if we keep them broad enough to include that which persists in organisms through material change.

EVOLUTION

In the eighteenth century, evolutionary theories built on the progressive narrative of the German Idealists. Reimagining the Neoplatonic hierarchy as a movement from simple to complex, biologists began to make sense of evidence that species have changed through time (e.g., fossils). The *scala naturae*, or ladder of nature, became an escalator.

Meanwhile, *gradualism* caught on in multiple fields, with the idea that dramatic changes could arise from millennia of slow, ubiquitous processes.[3] Scientists and natural philosophers began to favor theories of God as governor (eternal source of laws) over an interventionist God. In Biology, Charles Bonnet (1720–1793) and Jean-Baptiste Robinet (1735–1820) proposed that God created all organisms at once but reveals them only gradually. Past organisms and future organisms need not be the same. Their theories, however, still required a designer.

Comte de Buffon

An important shift occurred with George-Louis Leclerc, Comte de Buffon (1707–1788). Director of the French *Jardin du Roi*, he pioneered what we now think of as professional biology. Buffon described evolution without appeal to divine agency. With the organizing principle no longer held in the mind of God, students of biology could speculate once again about material, physical repositories for inherited order.

Along with many contemporaries, Buffon shifted attention from ontological categories to physical processes. He thought final causes were illusory and denied that vegetable and animal souls can be agents (Buffon 1797, Vol. 2, pp. 272, 347). Instead, we should think of the powers themselves as active. He identified the key vegetable powers as growth, expansion, reproduction, and increase (ibid., p. 260). All of these require the assimilation of new matter into the organism. Buffon believed in an infinite quantity of tiny organic particles, mixed in with the brute matter of the world. The process of nutrition separates the organic from the inorganic and incorporates particles into an organism according to an internal mould (*moule intérieur*). After growth is completed, additional organic atoms are sequestered as "seminal liquor" in the sex organs.

The internal mould does the same work that vegetable and animal souls did in the Middle Ages. It is not, however, an agent. The motivation comes instead from dynamic properties inherent in the organic particles (Sloan 2017). Buffon rejected the mechanical philosophy entirely (mechanical metaphor, ontological elimination, and etiological reduction), preferring to speak of organic particles with an inherent inclination

---

[3] Specifically, Uniformitarianism in geology and the Nebular Hypothesis in astronomy.

to order along the lines of Lucretius' soul seeds (Buffon 1797, Vol. 2, pp. 301–308).

He emphasized the fundamental character of nutrition and the continuity of vegetables and animals as one kind of life.

> This investigation induces us to conclude that there is no absolute essential and general difference between animals and vegetables, but that nature descends by degrees imperceptibly from an animal, which is the most perfect, to that which is least, and from the latter to the vegetable. (ibid., pp. 262–263)

Nonetheless, Buffon continued to see species as fundamentally static. The mould does not change through generations, though individuals may vary from it (Ruse 1996, p. 45). True origin of species would not be proposed until the next century.

Erasmus Darwin

A contemporary of Buffon, Erasmus Darwin (1731–1802) also emphasized the continuity of vegetables, animals, and humans. An English botanist and philosopher, he was the grandfather Charles Darwin. His book, *Zoonomia; Or the Laws of Organic Life* begins with a statement about the fundamental similarity of all life. It divides the world into two types of substances. Matter is entirely passive, while spirit is active; the latter moves the former. In plants and other living things, a "spirit of animation" drives biological processes and enables organization. Invisible to the senses, it nonetheless produces visible effects. The book also speaks of a subtle fluid, distilled from the blood by the pancreas and more refined than electricity (*Zoonomia*, 1:2). This fluid effects a *sensorium*, or inner-self, which permeates the body. Plants, with animals, possess all four faculties of the sensorium: irritability, sensation, volition, and association (1:5.2).

Erasmus Darwin remained committed to a physical account of vegetable and animal life, but he achieved this end by reframing the etiological reduction of the mechanical philosophers. He emphasized subjectivity among plants, placing it within the realm of physics (1:13).[4]

---

[4]Erasmus Darwin attributed ideas, love, dreams, and self-awareness to plants. Plants (and brutes) have volition, but humans are distinguished by deliberative and instrumental use of volition along the lines of Aquinas (*Zoonomia*, 1:11.2).

Thus, subjectivity may occur throughout the living world by the same principles. Along with Buffon, Darwin worked with an enhanced empiricism. His methodology did not require God as an efficient cause. Nor did it reject God as ultimate cause. It pragmatically focused on observable efficient causes and the continuity of humans with all life.

> I beg to be understood, that I do not wish to dispute about words, and am ready to allow, that the powers of gravity, specific attraction, electricity, magnetism, and even the spirit of animation, may consist of matter of a finer kind; and to believe, with St. Paul and Malbranch [sic], that the ultimate cause only of all motion is immaterial, that is God. St. Paul says, 'in him we live and move, and have our being;' and, in the 15th chapter to the Corinthians, distinguishes between the *psyche* or living spirit, and the *pneuma* or reviving spirit. By the words spirit of animation or sensorial power, I mean only that animal life, which mankind possesses in common with brutes, and in some degree even with vegetables, and leave the consideration of the immortal part of us, which is the object of religion, to those who treat of revelation. (1:14.1, Darwin 1796, p. 109)

This willingness to use the same, gradual rules to explain all aspects of life, including subjectivity, had a profound impact on his grandson and others. It also muddied the waters of empiricism. Should subjective life be allowed into the preserve of empiricism and natural science? If so, how have the lines drawn by the mechanical philosophers been redrawn?

Jean-Baptiste Lamarck

The first evolutionary biology with a genuine origin of species came from Lamarck, a French natural philosopher. Jean-Baptist de Monet, Chevalier de Lamarck (1744–1829) began his career as a botanist and, with support from Buffon, joined the growing node of biological research in Paris. Lamarck proposed a second mechanism of change in addition to the historical progress from simple to complex. Lamarck added the inheritance of acquired traits. Thus, the deer who stretches its neck repeatedly to reach leaves high on a tree will produce offspring with longer necks. He literally broke the mould of Buffon; the basic pattern of an organism was no longer fixed. It could be rewritten. Crucially, it could be rewritten by the interactions between organisms and their environment.

Lamarck described a process of slow change, wherein organisms develop gradually from simple to complex. All organisms differ from the inanimate world through "the faculties of alimentation [i.e., nutrition], development, and reproduction" (Lamarck 1914, p. 51). He tackled the motivation or efficient cause question directly. "Whatever may be the state of organisation [sic] of a body and of its essential fluids, active life could assuredly not exist in that body without a special cause capable of exciting its vital movements" (p. 186). That special cause comes in the form of subtle fluids, specifically the electric and caloric fluids (electricity and heat), present in the environment.

General forces were not sufficient to explain the individuality, unity, and operations of particular organisms. Lamarck completed his theory with a special force or factor that is more than the component molecules. Drawing on the *sensibilité* of the iatro-vitalists and the mould of Buffon, he spoke of a special kind of tension within the organism, which he called the *orgasm*. It echoed the life-breath of the Stoics. The orgasm unifies and empowers the organism. It differs from the *moule intérieur*, because it is not abstract and eternal. It changes with circumstance, allowing not just individuals, but species to evolve. This was the crucial step that allowed for new species to arise. Not just organisms, but the very types of organisms, which pass from parents to offspring, can change. Evolution brought life from simple forms to complex ones, from disorder to perfect order.

Lamarck thought plants and animals represented two distinct evolutionary trajectories. Animals have a dynamic ability to respond to the environment, which he called *irritability* and distinguished from orgasm. It resembles the soul-breath of the Stoics. This irritability has always been present in the animal lineage and marks the only true divide among organisms (pp. 51–60). Plants appear to have it, but that is only an illusion caused by mechanical interactions. Starting with spontaneous generation, the animal path runs from infusorians to mammals and the plant path runs from cryptogams to flowering plants.[5]

---

[5] Lamarck's categories are dated but correspond roughly to modern groups. Infusoria describes a group thought to be simple animals. Most of them have been classified as protists. Mammals remain the same. Cryptogams are plant-like organisms whose reproductive mechanism is hidden; they have no seeds. The group includes modern cyanobacteria, slime molds, fungi, algae, mosses, and ferns. Lamarck comments on our ignorance of plants, saying that we do not yet know them well enough to say which kind is most perfect, though clearly, it is among the flowering plants.

Like Erasmus Darwin, Lamarck favored strict physicalism but his idea of physicalism differed substantially from the mechanical philosophy. It was neither mechanical in metaphor nor reductionist in ontology. Instead, it followed the lead of the chemical philosophy, including subtle fluids as well as orgasm and irritability.

The eighteenth century witnessed several key advances in biological thinking. Biological order was moved out of the mind of God, allowing it to be both changeable and empirically knowable. Comte's epistemic confidence may have been too strong, but the willingness to explore what could be known physiologically proved essential to Charles Darwin and his accounts of animal and subjective life using vegetable activities: nutrition and reproduction.

Agency, which moved from God to the world with vitalist ideas, returned to life activities, acting within organisms or at the intersection of organism and environment. This completed a move back toward the in-action and in-fulfillment of Aristotelian souls. In this period, the *scala naturae* or chain of being shifted from an eternal expression of God's will, to a historical working out of God's will, and finally to a natural progression. Buffon, placed the process into the world of nature, while Lamarck created space for truly changing species. Charles Darwin would soon take the final steps to produce a fully non-progressive theory of evolution.

References

Buffon, Georges Louis Leclerc. "Barr's Buffon." In *Buffon's Natural History, Containing a Theory of the Earth, a General History of Man, of the Brute Creation, and of Vegetables, Minerals, Etc. From the French*, 10 vol., translated by J. S. Barr. London: H.D. Symonds, 1797.

Comte, Auguste. *The Positive Philosophy of Auguste Comte*, 2 vols. Translated by Harriet Marineau. Bristol, UK: Thoemmes Press, 2001.

Darwin, Erasmus. *Zoonomia; Or the Laws of Organic Life*, vol. 1, 2nd ed. London: J. Johnson, 1796.

Lamarck, Jean-Baptiste. *Zoological Philosophy: An Exposition with Regard to the Natural History of Animals*. Translated by Hugh Elliot. London: Macmillan, 1914.

Locke, John. *The Works of John Locke in Nine Volumes*, 12th ed. London: Rivington, 1824.

Ruse, Michael. *Monad to Man: The Concept of Progress in Evolutionary Biology.* Cambridge, MA: Harvard University Press, 1996.

Sloan, Phillip. "The Concept of Evolution to 1872." In *Stanford Encyclopedia of Philosophy*, Spring 2017 ed. Stanford University, 1997–. https://plato.stanford.edu/archives/spr2017/entries/evolution-to-1872/.

CHAPTER 16

Vegetable Significance: Evolution by Natural Selection

Charles Darwin (1809–1882) stripped evolution of its normative and progressive elements, setting the foundations for a new paradigm of life. He replaced the global movement toward complexity with a concrete mechanism for changing species: natural selection. He reintegrated formal and final causes, creating a new life-concept similar to Aristotle's nutritive soul. Natural selection provides a causal nexus, linking organization and purpose in organisms to a history of interactions between population and environment. Instead of projecting will and intention onto plants, Darwin allowed vegetable activities to explain animal life. He provided a physiological account, which emphasized vegetable proto-agency and made organization and purpose empirically tractable. Two contemporaries, Alfred Russel Wallace and Herbert Spencer, also defended evolution by natural selection. They retained more traditional (medieval) concepts of agency and progress, leaving their theories closer to vitalism.

CHARLES DARWIN

Darwin shared credit for the theory of natural selection with Alfred Russel Wallace, but he stood out for his empirical focus, his mechanical explanations, his marshaling of the argument, and his effective advocacy. Uninterested in becoming a physician like his grandfather Erasmus and father Robert, Charles showed early and continued interest in natural history. In college, he studied Divinity with an eye toward becoming an

© The Author(s) 2018
L. J. Mix, *Life Concepts from Aristotle to Darwin*,
https://doi.org/10.1007/978-3-319-96047-0_16

Anglican minister. This was a common path for natural historians at that time, providing both broad education and income. Raised by freethinkers, he first favored Unitarian theology. As life progressed, he moved toward deism and, eventually, a broad agnosticism.

At the beginning of his career, Darwin found Paley's design argument compelling. The order of living things requires an orderer. Darwin's great insight was to replace divine agency with *natural selection*, producing organization from the interactions of organism and environment.

Darwin was heavily influenced by Charles Lyell (1797–1895) and the rise of Uniformitarianism. The Uniformitarians claimed that extant observable causes were sufficient to explain geology. Gradual forces such as erosion and deposition, acting over great swaths of time, brought about valleys, mountains, and other features of the landscape. Darwin's gradual forces were reproduction and death, viewed as proximate efficient causes. Formal and final causes—particularly the final cause as ultimate, agential, or divine cause—were beyond human understanding. He argued that material and efficient causes were sufficient to explain the features of organisms.

In 1859, Darwin set forth his theory in *On the Origin of Species by Means of Natural Selection, or the Preservation of Favored Races in the Struggle for Life*. Populations of organisms always contain a variety of traits. Limited resources, such as food, mean that the environment cannot support all of the organisms in it. This results in a struggle for life. Many are born; few survive. When survivors can pass on successful traits to their offspring, those traits will increase in frequency within the population. This was well documented among domestic species. Breeders select long-eared dogs to breed and, as generations pass, the breeding population has longer and longer ears. Darwin suggested that nature acts as a breeder; hence, natural selection. No agency is required, however. Selection falls out of prolific reproduction and limited survival. As a result, the varieties that arise naturally will have been "optimized" for their environment.

Organisms do not require a designer as agent or efficient cause. Darwin still thought (at least at this time) that a cosmic *logos* was necessary. Natural selection works because of the presence of universal laws. Their origin and consistency require a creator and enforcer along the lines suggested by Bacon and Malebranche. And, with them, Darwin thought we could infer the existence of an orderer, but nothing about its character.[1] Biological accounts need no appeal to God beyond that needed for physics in general.

Throughout his career, Darwin remained publicly silent on two questions: God and the human soul. By the end of his life, his personal correspondence suggests a profound skepticism about both. And yet, he was careful to separate these questions from the concrete value of natural selection, which works as an empirical theory independent of these issues. In *Descent of Man* (1871), he compared the arrival of an immortal soul within a lineage of apes to the arrival of a human soul within an individual (Darwin 1882a, p. 613).[2] If the latter causes no anxiety, why should the former?

Darwin on Human Uniqueness

I am baffled by the common claim that Darwin reduced humans to animals. Humans were considered animals by most major thinkers in the Western tradition. Nor were theologians any less likely to call humans animals. The Aristotelian definition ("rational animal") emphasized that our animality was never in question; only our rationality. It was this rationality—along with subjectivity and spirituality—that was exported to Philo's invisible creation and Descartes' *res cogitans* realm. Human bodies were always animal. Human souls, on the other hand, had taken on all those aspects of humanity expelled from empirical consideration. Traditional soul concepts not only recognized, but strongly emphasized the extent to which embodiment constrains our rationality, subjectivity,

---

[1] "But I own that I cannot see, as plainly as others do, and as I should wish to do, evidence of design and beneficence on all sides of us. There seems to me too much misery in the world. I cannot persuade myself that a beneficent and omnipotent God would have designedly created the Ichneumonidæ with the express intention of their feeding within the living bodies of caterpillars, or that a cat should play with mice. Not believing this, I see no necessity in the belief that the eye was expressly designed. On the other hand, I cannot anyhow be contented to view this wonderful universe and especially the nature of man, and to conclude that everything is the result of brute force. I am inclined to look at everything as resulting from designed laws, with the details, whether good or bad, left to the working out of what we may call chance. Not that this notion at all satisfies me. I feel most deeply that the whole subject is too profound for the human intellect." Letter from Darwin to Asa Gray, 22 May 1860 (Darwin 2018).

[2] By the second edition, he has added a footnote citing the theology of James Picton who expatiates on the topic in *New Theories and the Old Faith*.

and spirituality. The animal and vegetable souls were slowly replaced by other terms more acceptable to the new scientific language (e.g., the *moule intérieur* and *orgasm*). Human souls remained an open question, if hermetically sealed from scientific study.

Within Darwin's lifetime, both natural theologians and reductive materialists were using "Darwinism" in support of their claims, but those claims predated Darwin and were based more on a priori ontology than on evolution by natural selection. T. H. Huxley (1825–1895) was a famous booster for materialists, but he appealed far more to Descartes than to Darwin in his famous essay on animals (including humans) as automata (Huxley 1874). On the opposite side of the debate, Charles Kingsley (1819–1875) and Asa Gray (1810–1888) argued that God upholds evolution. Their views were closer to Bacon and Malebranche. This is not to say that Darwin himself did not have an opinion. He did, and it was much closer to Huxley's. Nonetheless, the empirical portion of evolution remained bracketed from ontological questions of God and human souls—as Darwin maintained throughout his life.

Why then, was the new theory so controversial? I suspect it had something to do with Darwin's simultaneous identification of humans with animals and animals with vegetables. Throughout the long history of biology, both claims had been made, but almost never at the same time. Continuity was fine as long as it represented a flow: a Platonic condescension of *logos* or an Aristotelian ascension of matter. Darwin flattened out the whole process. He expressly linked vegetable activities to the development of both animal and human activities, specifically subjectivity. Darwin reduced humans to vegetables.

Darwin's works on the evolutionary foundations of subjectivity in humans are well known: *The Descent of Man* (1871) and *The Expression of Emotions in Man and Animals* (1872). His plant works are less well known, but equally revealing. *Insectivorous Plants* (1875), *The Power of Movement in Plants* (1880), and *The Movements and Habits of Climbing Plants* (1882b) all reveal an interest in the evolutionary foundations of sensation and motion, traditionally animal activities, within the plant kingdom.[3] Following his grandfather, Darwin saw all life as essentially

---

[3] Animal faculties are less transparent, but equally present in *On the Various Contrivances by Which British and Foreign Orchids Are Fertilized by Insects* (1862). From page 13: "I conclude that the action is not simply mechanical, but, for the want of a better term, may be called vital."

plant life. Natural selection allowed nutrition and reproduction to drive all other types of observable life, including sensation, locomotion, subjectivity, and the now reduced "reason."

Darwin made the claim explicitly in *Origin of Species*. "Analogy would lead me one step further, namely, to the belief that all animals and plants have descended from some one prototype" (Darwin 2009, p. 484). He aspired to a day when psychology would fall within the purview of evolutionary biology (p. 488). And all of this flows from the study of growth and reproduction.

Darwin not only replaced Aristotelian final causes with a historical chain of efficient causes, he replaced the Aristotelian first principle of biology (i.e., the threefold cause) with a new first principle (i.e., evolution by natural selection). Darwin reunited life as a common subject with a small set of explanatory principles. Darwin's approach was more consonant with Aristotle than it might, at first, appear, but it was utterly foreign to the "Aristotelian" psychology of the Middle Ages.

Darwin on the Environment as Cause

Before Lamarck and Darwin, students of life took for granted some permanence inherent in biology. They referred to the "fixity" of species or kinds. In the debate between Parmenides and Heraclitus, something must persist. It was cosmic *logos* for Plato and eternal form for Aristotle. The very notion of inheritance demanded that something pass from generation to generation. Ockham and Bacon called into question the existence of such categories, or at least our ability to know them. Descartes and Gassendi removed formal causes from the observable world but never truly eliminated them. Bacon, Leibniz, and Kant attempted to bridge the gap, but never truly succeeded. This left the field open for a range of descriptions running from the full-fledged vegetable soul of Stahl to the mould of Buffon. All of them treated biological species as fixed.

Darwin took a string of historical events—the interaction of populations with their environment—and used it to create a changeable organizing principle. Each new generation and each new environment leaves its mark on the population. Species still have an identity. They still pass on their traits to their offspring, but they pass on an ever-changing record of the past. And this, he said, is fully sufficient to explain the continuity of species. No agent is necessary, except the environment.

Such expressions as that famous one of Linnaeus, and which we often meet in a more or less concealed form, that the characters do not make the genus, but that the genus gives the characters, seem to imply that something more is included in our classification than mere resemblance. I believe that something more is included; and that propinquity of descent,—the only known cases of similarity in organic beings,—is the bond, hidden as it is by various degrees of modification, which is revealed to us by our classifications. (ibid., p. 413)

"Propinquity of descent" provided a causal nexus, as in Kant, whereby a final cause may be revealed as a string of historical efficient causes. No other final or formal cause is necessary. Darwin replaced the vegetable soul by invoking one cause, the historical relationship of organisms with their environment, as the efficient, formal, and final cause of organisms.

Strictly speaking, agents are still required: plants. Darwin's scheme rests on a population capable of nutrition and reproduction. Insofar as agency or proto-agency includes the assimilation of matter to form, it provides a basis for natural selection. The form has become more fluid, but it still acts on the environment, converting not-self into self. And the environment, in turn, acts on the form by natural selection in a population. Concepts of "individual" and "species" remain tendentious within biology for precisely this reason. Formal and final causes have been removed under their medieval definitions, but the nature of the causal nexus, the physics of nutrition, still does not fit neatly into the mechanical ontology.

Of course, natural selection is good enough to get on with. It does a tremendous amount of work without answering the ontological question. It tells us that populations change with time. Species come and go. It reflects and predicts how they change. It unifies diverse populations, explaining both similarities and differences. Looking back at vegetable and animal soul discussions, it explains the organization of living things and ties the activities of sensation, locomotion, and reason to a foundation of nutrition and reproduction. Perhaps one proto-agency is enough.

The epistemological question troubled Darwin, not the ontological one. He worried about God as final cause qua proximate efficient cause of organization and inheritance. Such an explanation was at best unknowable and at worst nonsense. "Propinquity of descent" did all the same work much more elegantly. Thus, he replaced vegetable and animal souls with a better causal account for individual organization and inheritance, one that required neither ontology nor metaphysics.

The details of material, physical inheritance still required the discrete genes of Gregor Mendel (1822–1884) and a century of molecular biology. By the mid-twentieth century, biologists identified deoxyribonucleic acids as the material locus of inheritance, with efficient cause explanations for both reproduction and variation. The language of vegetable and animal souls finally passed away, replaced by natural selection and genetics.

ALTERNATIVE EVOLUTIONS

Two contemporaries of Charles Darwin deserve mention. Despite broad agreement with him on the epistemological positions, they showed significant differences with regard to ontology. This breadth of biological interpretations contributed to philosophical differences that persist to the present day.

Herbert Spencer

One of the most prominent proponents of evolution, Herbert Spencer (1820–1903) defended an approach similar to Darwin's but still imbued with progressivism. A political and biological philosopher, he coined the phrase, "survival of the fittest." Spencer gave evolution a definite and evaluative trajectory, the improvement of species. He believed natural selection was inadequate for this and favored Lamarckian evolution (Weinstein 2017).

Michael Ruse spells out the position clearly. "[As] soon as we go beyond the fact of evolution, much of nineteenth-century British evolutionism and most of nineteenth-century American evolutionism owes far more to German transcendentalism than it does to anything in Darwin" (Ruse 1996, p. 181). Following Wolff, Goethe, and von Baer, Spencer saw a continuous motion in the universe—similar to the Platonic movement from becoming to being. It appears throughout history, from the formation of the solar system (nebular hypothesis) to the cooling of the earth, the evolution of species, and the succession of human races. At every point, Spencer argued for development from homogenous simplicity to heterogenous complexity (Spencer 1857).

In *Principles of Biology*, Spencer (1864) demonstrated a clear grasp of the material components of life as elements in the modern sense, specifically carbon, oxygen, hydrogen, and nitrogen (p. 3). He argued that life is more active than nonlife because organic molecules are highly mobile; oxygen, hydrogen, and nitrogen are all commonly encountered as gasses. Thus, the rarefied particles of antiquity transitioned to modern chemistry. He identified organization and nutrition as the defining features of life, thinking they require more than material explanation (secs. 1:4–5). And yet, he also denied any type of vitalism as the idea that "organic progress is a result of some in-dwelling tendency to develop, supernaturally impressed on living matter at the outset—some ever-acting constructive force, which, independently of other forces, moulds organisms into higher and higher forms" (pp. 403–404).[4] "We find progression to result, not from a special, inherent tendency of living bodies, but from a general average effect of their relations to surrounding agencies" (p. 430). Spencer emphasized the importance of interactions between organism and environment though with less attention to populations than Darwin (p. 80). Finally, Spencer saw no meaningful difference between plants and animals, though he considered it useful to have a different vocabulary for concepts of sensation and will (p. 96).

Alfred Russel Wallace

Darwin shared credit for natural selection with Alfred Russel Wallace (1823–1913). Younger than Darwin and less cautious in his epistemology, Wallace's reflections spurred Darwin to publish earlier than he might have, otherwise. Wallace increasingly turned to spiritualism as his life progressed. His accounts contained appeals to spirit and will, more in line with Erasmus Darwin than with Charles.

Wallace (1889, pp. 474–476) defended the traditional hierarchy of vegetable, animal, and rational life, arguing for three major transitions in the history of life. Each one represented an increase in complexity through the influence of an unseen world of spirit. The first transition, originating life or vitality, involved the formation of protoplasm, the life-specific matter, but also the activation of protoplasm by spirit. Living protoplasm is capable of nutrition, exemplified by carbon fixation,

---

[4]Spencer attributes this position to Erasmus Darwin, Jean-Baptiste Lamarck, Robert Chambers, and Richard Owen.

and reproduction: vegetable life. This life is capable of infinite variety.[5] The second transition, originating consciousness, involved biological complexity, but also the physically inexplicable phenomena of subjectivity: animal life. The third, based in complex consciousness, generated the unique and noble characteristics of humanity. The human faculties, including math, music, and art are not favored by adaptation and require some additional means to develop. Thus, human bodies evolved, but moral and rational abilities come from spirit (ibid., 478; see also Wallace 1870, Chap. 10).

Wallace complained that Huxley had embraced materialism too quickly. Protoplasm cannot *cause* organization; some order must have been present already for active protoplasm to arise. With echoes of Neoplatonism and Malebranche, Wallace argued that all of the natural forces, including gravity and electricity, depend critically upon some force of will. Humans, or some greater intelligence, must provide the motivation for all action in the universe (Wallace 1870, p. 366).

Importantly, this returned to the ontology/epistemology question. Darwin in middle age, like Malebranche, appealed to a fundamental order underlying all things, but considered it indiscernible from universal, physical order. Wallace reached for something more, something closer to intelligent design.

Once again, this view of life hinges on attempts to account for life under a mechanical paradigm without drawing the same boundaries as the traditional mechanists. Wallace (1910) defined life as a dynamic activity occurring within a physical substrate.

> Life is that power which, primarily from air and water and the substances dissolved therein, builds up organised and highly complex structures possessing definite forms and functions: these are preserved in a continuous state of decay and repair by internal circulation of fluids and gases; they reproduce their like, go through various phases of youth, maturity, and age, die, and quickly decompose into their constituent elements.

---

[5] "It has been well said that the first vegetable cell was a new thing in the world, possessing altogether new powers—that of extracting and fixing carbon from the carbon dioxide of the atmosphere, that of indefinite reproduction, and, still more marvelous, the power of variation and of reproducing those variations till endless complications of structure and varieties of form have been the result. Here then we have indications of a new power at work, which we may term vitality, since it gives to certain forms of matter all those characters and properties which constitute Life." Wallace (1889, p. 475, italics in the original).

They thus form continuous series of similar individuals; and, so long as external conditions render their existence possible, seem to possess a potential immortality. (pp. 3–4)

Wallace recognized the chemical components of protoplasm but felt that life also requires both "organizing power" and "directive agency" (pp. 358, 333).[6] The living world, without doubt, requires "some non-mechanical mind and power as its efficient cause" (p. 392).

In retrospect, Spencer's progressivism and Wallace's spiritualism appear radically different from Darwin's metaphysical agnosticism. At the time they were less dramatically opposed. Both demonstrate the bridge from earlier thinking to a physically oriented theory of life. Darwin's epistemic care proved immensely useful as the other positions faded into the background in biology.

Darwin provided a new foundation for biology by wrapping the efficient, formal, and final causes of organisms into one principle: natural selection. Starting with vegetable populations—that is populations capable of nutrition and reproduction—inheritance, variation, and limited resources lead, necessarily, to change. Looking at the interaction of organisms with their environment, we can see a causal nexus. Viewed chronologically, a long string of historical efficient causes shapes the forms of individuals and groups. Environments act on organisms. Viewed another way, organisms act on one another within populations, competing for resources. Organisms shape one another (and their environment). Their organization and purpose spring from the forms that, in turn, were shaped by efficient causes. Natural selection becomes an algorithm for optimizing nutrition and reproduction, so that one (perpetually changing) form can maximize its share of both the population and the environment.

In this way, natural selection was a vegetable soul. It differed from Aristotle's nutritive soul in two major ways. First, it did not partake of eternal forms. The forms of natural selection change constantly. Second, it depended radically on the interaction of environment and individual through populations. Darwin was able to take this step because of moves made by the early evolutionists in the previous chapters.

---

[6]Wallace listed these elements in order of decreasing abundance: H, C, O, N, S, Fe, P, Cl, Na, K, Ca, Mg. Modern lists resemble it closely (Mix 2009, pp. 76–82).

Final causes were naturalized (or efficient causes were broadened) to include interactions between organisms and their surroundings. Methodological commitments were softened and refined to allow for pragmatic descriptions of matter in motion that sidestepped both divinity and personhood. Formal and final causes were replaced by pragmatic, nominal analogs that could do the work of explanation. The last part of the book looks at modern biology and how vegetable and animal life-concepts have changed as a result of the new evolutionary paradigm.

REFERENCES

Darwin, Charles. *The Descent of Man, and Selection in Relation to Sex*, 1st ed. London: John Murray, 1871.

Darwin, Charles. *Insectivorous Plants*. London: John Murray, 1875.

Darwin, Charles. *The Power of Movement in Plants*. London: John Murray, 1880.

Darwin, Charles. *The Descent of Man, and Selection in Relation to Sex*, 2nd ed. London: John Murray, 1882a.

Darwin, Charles. *The Movements and Habits of Climbing Plants*. London: John Murray, 1882b.

Darwin, Charles. *The Annotated "Origin": A Facsimile of the First Edition of "On the Origin of Species"*. Annotated by James T. Costa. Cambridge, MA: Harvard University Press, 2009.

Darwin, Charles. "Letter no. 2814," Darwin Correspondence Project. Accessed 7 February 2018, http://www.darwinproject.ac.uk/DCP-LETT-2814.

Huxley, Thomas Henry. "On the Hypothesis that Animals Are Automata, and Its History." *The Fortnightly Review* 95(1874): 556–580.

Mix, Lucas J. *Life in Space: Astrobiology for Everyone*. Cambridge, MA: Harvard University Press, 2009.

Picton, J. Allanson. *New Theories and the Old Faith*. Edinburgh: Williams and Norgate, 1870.

Ruse, Michael. *Monad to Man: The Concept of Progress in Evolutionary Biology*. Cambridge, MA: Harvard University Press, 1996.

Spencer, Herbert. "Progress: Its Law and Cause." *Westminster Review* 67 (Apr 1857): 445–485.

Spencer, Herbert. *The Principles of Biology*, vol. 1. London: Herbert and Norgate, 1864.

Wallace, Alfred Russel. *Contributions to the Theory of Natural Selection: A Series of Essays*. New York: Macmillan, 1870.

Wallace, Alfred Russel. *Darwinism: An Exposition of the Theory of Natural Selection with Some of Its Applications*. New York: Macmillan, 1889.

Wallace, Alfred Russel. *The World of Life: A Manifestation of Creative Power, Directive Mind and Ultimate Purpose*. London: Chapman and Hall, 1910.

Weinstein, David. "Herbert Spencer." In *Stanford Encyclopedia of Philosophy*, Spring 2017 ed. Stanford University, 1997–. https://plato.stanford.edu/archives/spr2017/entries/spencer/.

PART IV

New Life

CHAPTER 17

"Vegetables" Versus Modern Plants

What remains of the vegetable soul? What can we learn from over two millennia of speculation on the meaning of life? Before we tackle these questions properly, we must resolve two common terminological confusions. Both result from the rapid changes in life-concepts over the past 300 years. Both arise at the intersection of history, philosophy, and biology. And both have been shaped by the discovery of dramatically different forms of life. This chapter covers the difference between the traditional vegetable, any living thing, and the modern plant, a member of the kingdom Plantae. The vegetable life category has expanded dramatically, while plants have become better understood and more narrowly defined. The next two chapters look at questions of individuality. Microscopic, chemical, and genetic investigations have revealed far more complexity than previously imagined. These new discoveries, along with a mechanical rejection of formal causes and Darwin's emphasis on populations, suggest a pragmatic approach to individuality. The final chapter ties up questions of agency and causation in biology with suggestions for how to proceed, taking the best from Aristotle and Darwin to provide a modern theory of nutrition.

© The Author(s) 2018
L. J. Mix, *Life Concepts from Aristotle to Darwin*,
https://doi.org/10.1007/978-3-319-96047-0_17

THE TREE OF LIFE

Darwin made one more important contribution, which I have not yet addressed. He promoted the idea of common descent: all organisms arise from one ancestor, or one ancestral population. By removing ideas of progress, he turned the ladder of life into a branching tree. As species diverged under natural selection, life became more diverse. There is not just one trajectory, but many. After Darwin, the terms "vegetable" and "plant" ceased to mean the lower portion of the *scala naturae* and started to reflect one branch among many. The tree of life metaphor had been prominent in the Middle Ages and Renaissance, but speciation by natural selection suggests that it is more than a strategy for taxonomy; it reflects the historical relationships between species. In the twentieth century, biological classification became almost exclusively *phylogenetic*, based on trees that reflect the course of evolution.

Modern taxonomy began with Carl Linnaeus (1707–1778), a Swedish physician, botanist, and zoologist. His *Systema Naturae* organized the natural world in tables of nested categories. For the narrowest categories he used *genus* and *species*, the now familiar "Latin binomial" (e.g., *Felis silvestris* for the domestic cat). They promoted international communication, replacing, or at least complimenting, vernacular common names. For the broadest categories, Linnaeus divided the Kingdom of Nature into minerals, vegetables, and animals, which came to be called kingdoms in their own rights. Minerals are aggregated bodies; Vegetables are organized bodies with life; Animals are organized bodies with life and feeling (Linnaeus 1758, p. 6). Later, he elaborated on animals, saying that they have sensation by means of nerves and motion by means of will (p. 8). Following tradition, Linnaeus included humans among the animals.

Biological classification slowly expanded, driven by the discovery of new species and new technologies (Margulis et al. 1994; Scamardella 1999). The mineral kingdom dropped from common usage, while plants and animals took on narrower definitions. From here on out, I will refer to them technically by their Latin designations (Regnum) Plantae and (Regnum) Animalia. I use "plant" to refer to the Plantae and keep "vegetable" for the broader, traditional life-concept.

The first major rearrangement came from the work of the Dutch microscopist, Antonie von Leeuwenhoek (1622–1723). A master lensmaker and observational biologist, Leeuwenhoek provided the first scientific record of microscopic organisms. In 1674, he reported these *animalcules* to the Royal Society in London. Over the next two centuries, biologists argued whether to classify them as primitive vegetables or primitive animals.

Three different biologists proposed adding a third kingdom in the 1860s. Richard Owen (1804–1892) proposed the Kingdom Protozoa for "early animals."[1] Owen also proposed more clearly mechanical descriptors for plants and animals. Plants, he said, are rooted, lack a digestive tract, form cellulose, and exhale oxygen. Animals have a digestive tract, inhale oxygen, and exhale carbonic acid. All other organisms were lumped into the Protozoa. John Hogg (1800–1869) proposed the Kingdom Primigenum (first born) or Protoctista (first created), suggesting they comprise the undifferentiated base of the tree of life, from which animals and plants emerged.[2] He objected to Protozoa, because it suggested that microorganisms are all animals. Ernst Haeckel (1834–1919) suggested Kingdom Protista (very first), including subgroups Protophyta (first plants), Protozoa (first animals), and Monera.[3] The last group, whose name means solitary or alone, included the smallest organisms, which had previously been considered by many to be stray cells, properly part of something larger. Some complained that the new kingdoms were defined by exclusion, what they are not, rather than positive characters, but no one knew enough to be more specific.

In the 1930s better technology, particularly the electron microscope, led to a second change. American biologist Herbert Copeland (1902–1968) noted differences between cells with nuclei (in Plantae, Animalia, and most known Protists) and those without (bacteria and blue-green algae, a.k.a cyanobacteria).[4] He took this as proof that the bacteria differ more from plants, animals, and protists than any of them differ from one another. Copeland proposed a new Kingdom Monera for organisms without nuclei, all unicellular. From the 1960s on, this group has been called prokaryotes (before nuclei), and other groups eukaryotes (true nuclei). Copeland went on to propose a much narrower definition of

---

[1] *Palaeontology*, 1860; The name protozoa had been suggested earlier by Georg Goldfuss, but not at the kingdom level.

[2] *On the Distinctions of a Plant and an Animal and on a Fourth Kingdom of Nature*, 1860.

[3] *General Morphology of the Organisms*, 1866.

[4] *The Kingdoms of Organisms*, 1938.

Plantae: organisms with chlorophyll a and b as well as xanthophyll and starch.[5]

Robert Whittaker (1920–1980) suggested yet another group, Kingdom Fungi on the basis of nutrition (Whittaker 1969).[6] The Fungi are multicellular eukaryotes with neither photosynthetic pigments (e.g., chlorophyll) nor a digestive tract. A five-kingdom system and attached story came to dominate biology. The oldest organisms were Monerans (i.e., prokaryotes). A eukaryotic lineage sprung up from within them, and three multicellular lineages then arose from within the eukaryotes: Plantae (or Metaphyta), Animalia (or Metazoa), and Fungi. The remainder of the eukaryotes remained lumped together, defined by exclusion as Protista (or Protoctista).

In the late twentieth century, molecular phylogenetics brought the third shift. A deeper understanding of nucleic acids as the material cause of inheritance reinforced phylogenetic thinking. New technologies for finding and comparing sequences led to much more detailed and sophisticated taxonomy. American geneticists Carl Woese and George Fox (1977) discovered they could construct a master tree of life using small-subunit ribosomal RNA, a string of nucleic acids highly conserved across all known species. The new tree revealed surprising diversity among the prokaryotes. A previously unknown group of one-celled organisms, the Archaebacteria (or Archaea), share some traits with bacteria and some with eukaryotes. This suggested a new system, still current in biology. Life has three *domains*: Eubacteria (true bacteria, including most known prokaryotes), Archaebacteria (ancient bacteria, including many extremophiles), and Eukaryota (organisms whose cells have nuclei).

Viewing this as a tree of life, growing in time, the three domains diverge somewhere near the root, the origin of life. Each branches out into a wide array of living beings. Plantae forms one of perhaps six major branches within the Eukaryotes. Animalia and Fungi, more closely related to one other than to any other group, form a second branch. The Archaebacteria have three branches, still only poorly characterized, while the Eubacteria may have more than twelve. One-celled prokaryotes lack

---

[5] *The Classification of Lower Organisms*, 1956.

[6] Similar suggestions were made by Lynn Margulis at the same time.

the structural diversity of Plantae, Animalia, and Fungi, but they have far greater biochemical and genetic diversity.[7]

MODERN PLANTS

The modern group, Plantae includes most organisms commonly thought of as plants today: grasses, ferns, bushes, trees... Formally, it describes a group of highly evolved, multicellular organisms with nuclei and two types of symbiotic bacteria: mitochondria and chloroplasts (Mix 2009, pp. 242–243, 254–257). The mitochondria, present in all eukaryotes, convert the products of sugar metabolism into usable chemical energy. Mitochondria may be necessary for plant, animal, and fungal life as we know it. They provide far more efficient energy usage than other chemical processes, and it is difficult to imagine multicellular life without them (Smith et al. 2016). Chloroplasts, present in plants and a number of other organisms, contain vast arrays of photosynthetic pigments. They use solar energy to assemble sugars. Plants also commonly possess cell walls with cellulose and reproduce sexually, alternating between haploid and diploid generations.

Mitochondria and chloroplasts answer, at least in part, Plato's question of motivation. They take energy from the environment and turn it into the chemical energy that powers cells. Chloroplasts convert sunlight into chemical energy. Plants and other *autotrophs*, or self-feeding organisms, build complex molecules, including sugars to store the energy for later use. Mitochondria break those sugars down again, producing more easily used chemical energy. Mitochondria and chloroplasts motivate many organisms. They could not have arisen, however, without first being organisms themselves. Phylogenetic analysis shows that both were once free-living bacteria. Now they exist only within the cells of eukaryotes.[8] Significantly, they still reproduce and, if lost, cannot be replaced by their host.

---

[7]The definition of "multicellular" remains troubling. Many bacteria form colonies that cooperate for nutrition and reproduction, but do not have the obligate cooperation between cells we see in the Plantae and Animalia.

[8]Mitochondria arose within the proteobacteria, while chloroplasts are closely related to the cyanobacteria. A few plants without chloroplasts exist, but they are clearly descended from more usual plants.

The traditional category of vegetable life should not be confused with the group Plantae. The vegetable life-concept embraces everything with nutrition and reproduction—in other words, all life. Plantae, Animalia, and Fungi are all groups of vegetables with extraordinary complexity and function. Each can be thought of as one evolutionary trajectory. Each displays multiple, complex, adaptive features, including both structure and behavior.

Sensation in Plants

Plant complexity extends to many traits long associated with animals. The conceptual foundations of sensation, imagination, and will remain problematic empirically. Still, botanists recognize clear Plantae analogs. Environmental conditions become biochemical signals, get stored, and change how the plant interacts with its surroundings.

Easiest to study empirically, the front end of sensation involves signal-formation. In humans, touch, taste, smell, hearing, and vision all convert environmental factors into electrical signals in the nerves. Nerves conduct the signal to the brain. The qualia of sensation, the subjective experience of these signals by a conscious mind, what I call the back end of sensation, remain mysterious, but the front end can be seen in some members of the Plantae as well as in the Animalia. Plants have neither nerves nor brains, but many have alternative mechanisms for passing and recording signals.

The Venus Flytrap provides a classic example of plant touch. Growing in nitrogen-poor soils, this plant captures insects as nutritional supplements. Two strong leaves lie flat, connected by a hinge. A fly stepping onto the leaves triggers a response. The leaves snap closed, trapping the fly. The plant then dissolves it and absorbs the material. Wind and water do not trigger the trap. Venus Flytraps achieve this specificity with sensory hairs on the leaves. Pressure on the hairs produces an electrical potential, but only certain combinations of pressure spring the trap. Similar responses to pressure have been observed in the shrub *Mimosa pudica*. A cocklebur (*Xanthium strumarium*) and a mustard (*Arabidopsis thaliana*) change their growth pattern if their leaves are touched. Oak, beech, and spruce leaves all produce both electrical and chemical signals when eaten by insects (Wohlleben 2016, pp. 7–8). Botanists are just starting to look at plant touch, but both electrical and chemical signals occur as a result of variations in pressure and temperature—just as in animals.

The related senses of taste and smell occur when organisms react to chemical signals in the environment. Consuela De Moraes, a Biologist in Zurich, studies olfactory cues in plants (Chamovitz 2012, pp. 39–43). She has shown that the five-angled dodder plant (*Cuscuta pentagona*) smells its neighbors. Like the Venus Flytrap, nutritional challenges make all the difference. A plant vampire, the dodder lacks both roots and leaves; it siphons off water and nutrients from other plants. Chemical signals in the air reveal the plants which surround it. Placed between a tomato plant and wheat, the dodder "sniffs out" the most attractive victim and grows toward it. Countless other examples of taste and smell have been documented, from a simple response to ethylene in the air (which causes fruit to ripen more quickly), to intraspecies signaling (releasing ethylene to signal distress), to detection of predators (Wohlleben 2016, pp. 6–13; Chamovitz 2012, pp. 33–59; Helms et al. 2017). Scientists have documented light detection ("sight") and sound detection ("hearing") as well (Chamovitz 2012, pp. 9–31, 87–110; Wohlleben 2016, pp. 148–149).

Memory and Will in Plants

The front end of sensation occurs in plants. What about the back-end? And, can plants turn the sensed signals into willed action? The objective location of "images" remains controversial, but plant activities do change in response to environmental changes. They make "choices" based on the past conditions (Chamovitz 2012, pp. 141–165). Environmental cues impact gene expression, causing genes and gene clusters to turn on or off depending on local conditions. Such states can even persist through multiple generations.[9] So, a plant may remember an environmental cue and act on it much later in time. For instance, some plants conserve resources in response to a bad winter, even after years of more clement weather. No image or internal representation would be necessary for such a process, but it does provide a weak analog for memory and will.

---

[9] Technically, this is epigenetics and not inheritance. Although a trait persists through multiple generations, the underlying genes remain the same. An individual gene codes for *conditional* expression, based on the environmental cues, even when those environmental cues are distant in time. Thus, this is not Lamarckian inheritance of acquired traits. Genetic variation, inheritance, and selection still follow natural selection.

If memory and will are nothing more than signal-mediated responses to environmental conditions, then many plants possess this as well as animals. Many "simpler" organisms do as well. If, on the other hand, they require something more (perhaps subjectivity, internal representation, or the ability to anchor a causal chain) then we must open a discussion as to how such traits would be verified. It is not obvious that such distinctions are amenable to empirical study within the paradigm of modern biology.

The traditional concepts of animal life and animal soul apply to organisms with sensation and locomotion. These activities present challenges within the mechanical paradigm. If we accept a priori that sensation and locomotion are nothing but mechanism, then we must say that some representatives of the Plantae possess animal life (along with many other organisms). If, on the other hand, we claim that the interior life of animals is beyond observation, then we must admit that it might also be beyond observation in plants. The one thing we cannot do is a claim that modern science has discovered a difference between the two groups along these lines. To be clear, we can say that Animalia and Plantae are distinct groups. Chloroplasts and cell walls are just two of the very clear differences. We cannot say they are distinct with respect to sensation and locomotion, the traditional hallmarks of animal life.

THE VALUE OF VEGETABLE LIFE

Contemporary scholars in a variety of fields have questioned the traditional preference for animals over plants (e.g., Hall 2011; Marder 2014; Nealon 2015; Marder and Irigaray 2016). Though they provide an important critique of human-centered (and Animalia centered) thinking, they often conflate traditional vegetables with the modern Plantae. Normative discussions about life, and the relevant value of various kinds, depend on philosophical context and the categories we use.

A first question relates to life versus nonlife. Across time, there seems to be a consensus that something strange and interesting occurs in all beings engaged in nutrition and reproduction. The traditional vegetable life-concept refers to this kind of activity and this kind of being. Scholars have debated whether vegetables should be thought of as alive and ensouled; they have not, for the most part, debated whether they constitute a clear category.

A second question relates to vegetable versus animal life. Traditional thinking, both Aristotelian and Platonic, viewed animals as a special kind of vegetable. Animals had more value because animals could do everything vegetables could do (vegetable faculties) plus more (animal faculties). One could argue that the distinction should not make a difference, that there is nothing inherently valuable about sensation and locomotion, but it would make no sense to argue that vegetables are valuable because they are equally powerful or complex. In the traditional scheme, they are not—by definition. Recent discoveries only reveal that animal life is more common and diverse than previously realized.

On the other hand, we might consider the life of Plantae, versus the life of Animalia. Both are complex in their own way. As are Fungi. Multiple successful lineages have arisen that succeed with complex bodies, sensation, and greater proto-agency than their relatives. Within the Animalia, this takes one form; within the Plantae, another; in Fungi yet another. No one of these groups is more advanced, complex, or evolved by biological standards. Not all specialization is progress toward a single goal.

Looking closely, I might further argue that bacteria and protists are equally evolved. Animalia, Plantae, and Fungi have succeeded through growing large multicellular bodies. Other organisms have succeeded in other ways. The cyanobacteria, for instance, remain "simple" in that they do not grow multicellular bodies like the Plantae, but they have expanded into almost every environment on Earth, including the cells of all plants. They have sophisticated structures inside their cells for capturing sunlight and processing chemicals. They are also responsible for altering the chemistry of Earth's atmosphere and hydrosphere to an extent unparalleled even by humans (Mix 2009, pp. 175–176). The oxygen they produced poisoned all other forms of life on the planet and made multicellular life possible.

With an evolutionary life-concept, there is no higher and lower life. There are only different strategies and different trajectories. If anything, the critics of animal-centrism do not go far enough in recognizing the multitude of ways that life can be valuable. Complexity in the Plantae says nothing about vegetable life, or vice versa. Such comparisons are anachronistic.

In rejecting progress, the evolutionary life-concept must also reject "higher" and "lower" organisms. All leaves on the tree of life, all living organisms, are the same distance from the root, some three-plus billion years in the past. Many organisms and lineages have discovered unique and sophisticated ways of dealing with their environment. Indeed, just such adaptations prove successful and lead to a proliferation of offspring. Lineages survive because they have successful traits. The Plantae and Animalia are two such lineages. Plantae, the modern plants, are many-celled eukaryotes with cell walls and chloroplasts, domesticated bacteria that use chlorophyll to convert sunlight to energy. Animalia, the modern animals, are many-celled eukaryotes without cell walls that depend on other organisms for their food. They commonly move and ingest their food. Numerous other such lineages exist, including the Fungi and cyanobacteria.

The traditional vegetable life-concept corresponds to all life, including the Plantae and as well as Fungi, cyanobacteria, and everything else. If we measure only the mechanical aspects of sensation and loco-motion, the front end, then many of these groups fit the traditional ani-mal life-concept as well as Animalia. They receive, store, and "act" on signals from their environment. The subjective life-concept along with associated qualia and more modern definitions of will and agency remain beyond the scope of scientific epistemology, as do spiritual and rational life-concepts.

Plants just aren't what they used to be. Understanding both the mod-ern category and the traditional life-concept will require differentiating the two. Both remain useful, but the traditional life-concept now applies to all life. Our understanding will also hang on our ideas of inside and outside, discussed in the next chapter.

REFERENCES

Chamovitz, Daniel. *What a Plant Knows: A Field Guide to the Senses of Your Garden—And Beyond*. London: Farrar, Straus and Giroux, 2012.

Hall, Matthew. *Plants as Persons: A Philosophical Botany*. Albany, NY: SUNY Press, 2011.

Helms, Anjel M., Consuelo M. De Moraes, Armin Tröger, Hans T. Alborn, Wittko Francke, John F. Tooker, and Mark C. Mescher. "Identification of an Insect-Produced Olfactory Cue that Primes Plant Defenses." *Nature Communications* 8 (Published online 2017). https://doi.org/10.1038/s41467-017-00335-8.

Linnaeus, Carolus. *Systema Naturae*. Stockholm: Laurentius Salvus, 1758.

Marder, Michael. *The Philosopher's Plant: An Intellectual Herbarium*. New York: Columbia University Press, 2014.

Marder, Michael, and Luce Irigaray. *Through Vegetal Being: Two Philosophical Perspectives*. New York: Columbia University Press, 2016.

Margulis, Lynn, Karlene V. Schwartz, and Michael Dolan. *The Illustrated Five Kingdoms: A Guide to the Diversity of Life on Earth*. New York: HarperCollins College, 1994.

Mix, Lucas J. *Life in Space: Astrobiology for Everyone*. Cambridge, MA: Harvard University Press, 2009.

Nealon, Jeffrey T. *Plant Theory: Biopower and Vegetable Life*. Redwood City, CA: Stanford University Press, 2015.

Scamardella, Joseph M. "Not Plants or Animals: A Brief History of the Origin of Kingdoms Protozoa, Protista and Protoctista." *International Microbiology* 2, no. 4 (1999): 207–216.

Smith, Felisa A., Jonathan L. Payne, Noel A. Heim, Meghan A. Balk, Seth Finnegan, Michał Kowalewski, and S. Kathleen Lyons, et al. "Body Size Evolution Across the Geozoic." *Annual Review of Earth and Planetary Sciences* 44 (2016): 523–553.

Whittaker, Robert. "New Concepts of Kingdoms and Organisms." *Science* 163, no. 3863 (1969): 150–160.

Woese, Carl R., and George E. Fox. "Phylogenetic Structure of the Prokaryotic Domain: The Primary Kingdoms." *Proceedings of the National Academy of Sciences, USA* 74, no. 11 (1977): 5088–5090.

Wohlleben, Peter. *The Hidden Life of Trees: What They Feel, How They Communicate—Discoveries from a Secret World*. Vancouver, BC: Greystone Books, 2016.

CHAPTER 18

Vegetable Individuality: The Organismal Self

Within the new biological paradigm, evolutionary and broadly mechanical, twentieth-century biologists struggled to clarify a new life-concept that included the variety of known life. As early as Plato, living bodies were seen to be more than physical aggregates of matter. They cohere. Any lump of carbon constitutes a physical body; few constitute a living body. Even fossils require a biological explanation for their order. In chapter one, I spoke about a Martian meteorite. We care about whether that meteorite has biological bodies in it. And, we think there are ways to tell the difference.

Traditional life-concepts take animal unity and individuality for granted but there was often debate about vegetables. Animal souls produced discrete animal individuals. Vegetable souls could do the same, but vegetable natures might not. By explaining animals with vegetable faculties, Darwin called their unity and individuality into question. With an evolutionary life-concept, what should we make of organismal unity? I argue in favor of *biological nominalism*. Biology as natural science should treat individuals (organisms, species, units of selection, and units of inheritance) as arbitrary names, useful to the extent that they allow us to make accurate predictions within the framework of evolutionary theory. Key to biological nominalism will be the idea that claims of organismality are not exclusive. One organism may overlap another. This represents a departure from traditional life-concepts precisely in our use of formal causes.

© The Author(s) 2018
L. J. Mix, *Life Concepts from Aristotle to Darwin*,
https://doi.org/10.1007/978-3-319-96047-0_18

The twentieth century began with multiple attempts at an expressly scientific definition of life. The top contenders came from updated versions of nutrition and reproduction (Mix 2014, 2015). Nutrition became self-regulation, focused on individuals that maintain internal conditions in a changing environment. Reproduction became self-replication, focused on individuals that perpetuate themselves.

REGULATORS

The word 'metabolism' appeared near the end of the nineteenth century to describe the full set of chemical reactions involved in growth and life, specifically building up and breaking down complex carbon chains. French physiologist Claude Bernard (1813–1878) pioneered metabolic research and suggested an internal environment (*milieu intérieur*) for organisms (Cooper 2008). Bernard divided living things into three categories. *Latent life* lacks visible metabolism. *Oscillating life*, including plants and most animals, varies depending on the environment. Metabolism only becomes apparent under certain external conditions. Only *constant* or *free life*, which he identified with warm-blooded animals, manifests metabolism at all times. The ability to maintain a stable internal environment makes free life the highest and best form of life. J.S. Haldane (1860–1936) expanded on this, saying that "Biology must take as its fundamental working hypothesis the assumption that the organic identity of a living organism actively maintains itself in the midst of changing external appearances" (J.S. Haldane, *Respiration*, 391, quoted by Cooper 2008, p. 423). Debate persisted about the level of reductionism and agency inherent in these remarks, but the idea caught on in biology and medicine. Walter Cannon (1871–1945) coined the term *homeostasis* for physiological regulation of the internal environment in a stable state. He spoke of "agency," but saw it as automatic and non-intentional. Later, John von Neumann (1903–1957) developed ideas of feedback loops and regulatory systems, comparing artificial and living systems.

Bernard described self-regulation but focused on "higher" animals. Early twentieth-century thinkers proposed it as a broader definition of life and as a boundary condition for *singular* lives. Soviet biochemist Alexander Oparin (1894–1980) and British biologist J.B.S. Haldane (1892–1964) suggested that metabolic processes define life (Bernal 1967; Haldane 1929). Both distinguished between the early Earth and the modern Earth. Energy was once abundant and sufficient to stimulate spontaneous generation; now it is not. Energy caught up in metabolic systems allows them to behave differently than their environment. The Austrian physicist Erwin Schrödinger (1887–1961) famously defined life thermodynamically, replacing chemical self-regulation with energetic self-regulation (Schrödinger 1992). He claimed that life avoids thermodynamic equilibrium by "feeding" on "negative entropy." In other words, life eats order to maintain order. Whether one considers the key feature of life to be chemical or energetic, these theories give us a picture of organisms persisting through changing material by absorbing resources. An organism is regulator, regulated system, and the process of regulation.

By the late twentieth century, mechanical metaphors returned, partially in response to the development of computers and partially because the old discussions had passed out of popular consciousness. Jacques Monod (1972), Ernst Mayr (1961, 1982), Freeman Dyson (1985), and Daniel Dennett (1995) all suggest biological programs or algorithms running in a material processor. Programs direct and motivate biological regulation. Since the Aristotelian Synthesis, "nutrition" has included not only assimilation of food, but the exchange of material within a living body. Concepts of homeostasis and metabolism perpetuate this idea within the framework of modern biology.

REPLICATORS

A second school of thought followed reproduction. Once the stability of species moved out of the mind of God, it became necessary to find a physical mediator for inheritance. A new, mechanical biology required a physical repository of inherited order and some proto-agency to copy it. Darwin's theory of natural selection lacked a mechanism for inheritance.[1] Gregor Mendel (1822–1884) provided a mathematical treatment of discrete hereditary units or genes. In the first half of the twentieth century, the material components of genes were localized, first to the nucleus, then to nucleic acids, and eventually identified as sequences within long chains of nucleotides. James Watson and Francis Crick provided a

---

[1] Darwin proposed the transmission of *gemmules*, small particles given off by organs and concentrated in the gametes.

physical mechanism with their groundbreaking paper on the structure of DNA in (1953).

Early twentieth-century biologists also began to explore the possibility that protoplasm—the material of life—could be explained with an appeal to some minimal, replicating unit. Dutch geneticist Arend Hagedoorn (c. 1885–1953) spoke of molecules capable of copying other molecules and even copying themselves (Hagedoorn 1911). In the United States, his theory of *autocatalysis* was picked up by physicist Leonard Troland (1889–1932) and geneticist Herman Joseph Muller (1890–1967), who specifically tied his theories to nucleic acids (Troland 1917; Muller 1926, 1966). Replicator theories of life focus either on genes, viewed as self-replicating molecules, or some larger biological entity that copies itself, using information in the genes. Some have expanded this basic concept to include different units of inheritance, such as lines of code or memes.

Unlike self-regulation, self-replication is not a continuous process. This can make it difficult to speak of replicators in the present tense. It is easy to identify a string of replicators through time, but beginnings and endings cause problems. How could a first replicator arise? Are the products of replication alive, even when they cannot replicate? Intuition suggests that mules are alive, even though sterile, but unfertilized eggs are not. And, what are we to make of entities that might replicate, but have not been observed doing so?

These challenges led to two common addenda to replicator theories of life: potentiality and self-regulation. Some biologists speak of organisms as "potential replicators" or "capable of replication." Others include some aspect of chemistry, metabolism, or cellularity to limit the idea to familiar examples. One of the most popular modern life-concepts, the "NASA Definition," includes both addenda: "life is a self-sustained chemical system capable of undergoing Darwinian evolution."[2] This duality captures a desire to unify replicator and regulator concepts in a single theory.

---

[2] Joyce (1994) provides the words, saying only that they arose in discussion during a NASA Exobiology Program meeting. Luisi (1998) and Benner (2010) started calling it the "NASA Definition." NASA has never officially adopted it, though the Astrobiology Programs use it as an operational definition for planning purposes. It closely follows Carl Sagan's genetic definition in "Life" for *Encyclopedia Britannica*.

BIOLOGICAL INDIVIDUALS

Neither regulators nor replicators reliably align with our intuitions about what it means to be an organism. Neither do they reliably align with one another. In this context, I will use the language of *ramets* and *genets* to clarify the differences. A ramet has physiological unity. It names a regulatory individual, especially when that individual may not be a distinct "organism." A genet has genetic unity. It names a replicatory individual, especially when it may not be an "organism."

In bacterial colonies, one genet includes many ramets. A single cell has cloned itself, producing a group of genetically identical units. The phenomenon also occurs among Protists, Fungi, Plantae, and even Animalia (Godfrey-Smith 2009, pp. 70–81). A remarkable example can be seen in the quaking aspen named "Pando" (DeWoody et al. 2008). Tens of thousands of ramets, "trees" with distinct trunks, spread out over a hundred acres in Utah. They have identical genomes and, thus, identical genetic interests under natural selection. This makes them one genet.

Conversely, we can consider fungal hyphae. These long, branching chains of cells have perforated walls allowing them to share nutrients, cellular machinery, and even nuclei. It was long assumed that the many nuclei within a single hypha shared identical genomes. Recent studies show that this is not always true (Johannesson and Stenlid 2004; Roper et al. 2011). It is possible for a single hypha, as ramet or regulatory unit, to contain multiple genets.

Perspectives began to shift in the mid-twentieth century with the discovery of intracellular endosymbiosis, one organism living inside the cells of another (Sagan 1967). Both mitochondria and chloroplasts, essential to our current understandings of eukaryotes and the Plantae, appear to be bacteria living inside eukaryotic cells. These bacteria have their own nucleic acids and replicate themselves. They have their own metabolism and regulate themselves, out of equilibrium with the chemistry of the surrounding cell. They cannot live without their host. Their host cannot live without them. Because they are essential to the nutrition and reproduction of their host, they are "inside" their host organically as well as spatially: one organism within another.

Stranger still, some ramets partially overlap with genets (Godfrey-Smith 2009, p. 75). Marmosets and Tamarins, small South American monkeys, produce fraternal twins at an unusually high rate. Two individuals with two different genomes develop within their mother's body. The twin embryos swap cells, becoming *chimeras*. Different tissues within the same body have different genomes, different genetic templates. They even swap reproductive tissue. This means that either body can pass on either genome to their offspring. Two ramets, two genets, but the boundaries occur in different places. Which boundaries define individuality? Which define an organism?

Modern evolutionary biology abounds with complicating examples. Genes can compete within a genome (selfish genetic elements, Dawkins 1976). They can be transmitted independently of bodily reproduction (horizontal gene transfer). Cells can compete within a body and even become parasites in other bodies (transmissible cancers, Ostrander et al. 2016). When every Eukaryote—including Plantae, Animalia, Fungi, and every traditional vegetable and animal—involves a symbiosis of multiple species, symbiosis is the norm, not the exception.

Genetic Individuality and Agency

From this perspective, it might be reasonable to adopt a selfish gene mentality. In *Selfish Gene*, Richard Dawkins advocates thinking about genes as the true units of biology, with bodies (the traditional individuals) acting only as vehicles. Such genes act as the proto-agents, using cells, bodies, and even populations as their tools. This approach has important advantages. It allows clear language for thinking about how natural selection can act differently on each of the many genes within a body. And yet, "genes" are abstract entities subject to the same critiques leveled at vegetable souls. Have they become inappropriately substantial, agential, or intentional? Surely, proponents use the ideas metaphorically and analogically. But, have they used them on genes more carefully than their predecessors used them on souls, spirits, moulds, and the like? Success will require attention to detail.

At first blush, the gene seems like a straightforward entity, a string of nucleotides that provides the basis for inheritance. The gene persists through time and changing material, however, much like an organism. Nucleotides can be replaced without losing genetic identity. Natural selection works on gene *types*, the collection of nearly identical gene copies distributed throughout a population.[3] Sometimes, a gene may be adapted to benefit the type, even when one physical instantiation (a *token*) harms the ramet in which it resides. David Haig notes, "A gene that is distributed throughout the world cannot be part of an organism that is localized in space" (Haig 2012, p. 469). Haig is quick to point out that this is not an insoluble problem. It simply requires us to distinguish the local, material and physical gene token from the abstract, distributed gene type. Dawkins' selfish gene refers to the latter. In some ways this is a return to a Platonic physics, in which we speak of agency at the level of universals.

Numerous philosophical issues arise. Mary Midgley (1985, pp. 143–154) objects to the selfish gene metaphor, starting with the attribution of agency and moving on to the attribution of motives, hidden assumptions, and hasty generalization. Peter Godfrey-Smith (2009, p. 143) argues against selfish gene language, saying that "these formulations... became more than a shorthand, being used not just to summarize complicated ideas but to shape the foundational descriptions of evolution." The selfish gene goes beyond natural selection as causal nexus. It creates a new ontology that invokes agency for genes while denying agency to anything else.

This perspective is neither necessary nor universal among biologists, but it has become common. Evelyn Fox Keller (2002), in a more descriptive vein, speaks about the development of the term 'gene' in a way that hides, or at least condenses, the bridge between physics and biology that was once covered by the vegetable soul. Citing Arnold Ravin, she asks how to reconcile material heredity with the power to cause replication and direct development.

> [The gene] melds into a single form two entities with the disparate properties of atom and organism, and contains the incoherence of such a melding under the protective wrap of a new word. In effect, it offers a resolution of the riddle of life by invoking an entity that is riddle in and of itself. (p. 131)

Keller goes on to describe the shift in metaphor from gene to genetic program and the challenges of attributing agency there as well (p. 136).

I do not wish to deemphasize the tremendous importance of gene-centered investigations. William Hamilton (1936–2000) and others have shown how often we must consider natural selection acting on

---

[3]The type/token distinction is borrowed from Millikan (1984).

genes. Dawkins' work and popularization of this insight has been valuable. In order to avoid the traps of previous life-concepts, however, these theories must not privilege genes by turning them into essences or agents in an exclusive way. Haig (2012, p. 468) quotes Godfrey-Smith (2009, p. 136) in saying that the way we speak of genes is necessarily pragmatic and somewhat arbitrary. "In an evolutionary context, it is more accurate to talk of genetic material, which comes in smaller and larger chunks, all of which may be passed on and which have various causal roles." With these authors, I believe that neither gene nor organism is entirely arbitrary. Both reflect real phenomena. Both are useful precisely when they catch the causal nexus described by Kant and Darwin. The key insight, however, is that they do so in a non-exclusive way. Multiple, non-compatible conceptions of individuality may usefully track reality. Thus, biology as a discipline can include multiple research programs, each using a different theory of gene and organism. No essences or natural kinds are needed. Indeed, invoking them prevents alternate research programs without adding clarity.

If genes as replicators act as agents, other replicators can as well. If genes compete with other genes within a genome, then the genome can be considered population and environment as much as the more familiar extra-cellular or extra-organismic milieu. We must not fall into the trap of thinking that genes are metaphysically or causally unique entities.

Souls were used to provide biology with individual unities, largely through the use of the soul as the formal cause or essence of an organism. Such formal causes do not fit into the natural sciences, being removed from the extended world by the mechanical philosophers. Bacon, Leibniz, and Kant all tried to put them back into "nature," but ended up rejecting any way to discern them empirically. Individual organisms may be allowable pragmatically and nominally, but there is no reason to allow them ontologically. In the words of Nietzsche, "there is no form in nature, for there is no inner and no outer" (quoted in Miller 2002, p. 152).

BIOLOGICAL NOMINALISM

Vegetable souls provide a scaffolding on which to construct a new life-concept that avoids the pitfalls of previous attempts, particularly with regard to biological agency and individuality. A causal nexus, joining efficient and final causes can be used to assign names to biological lineages.

This follows Aristotle in defining life (and biological explanation) by a confluence of efficient, formal, and final causes. It departs from him, and even more from his medieval interpreters, by making formal causes historically contingent and non-exclusive. One concept of biological individuals can embrace genes, classical organisms (as genet or ramet), and species without contradiction.

Subsistent and eternal forms have been lost; nominal forms remain. The subsistent form formal cause rejected by Bacon remains off limits, but alternatives exist. We need not choose between natural kinds and totally arbitrary names. Some kinds are more natural than others. More to the point, some labels lead to more useful predictions (Millikan 1984; Shields 2012).

It has become popular to claim that life lacks clear boundaries. Shields (2012) argues that "life" is a core-dependent homonym with no essential traits, but related to intrinsic ends. Godfrey-Smith (2009) speaks of groups that range from *minimal Darwinian* populations to *paradigm Darwinian* populations, with only the latter being clear examples of life. Many in the astrobiology and origin-of-life communities hope that a fuzzy definition will make it easier to discover intermediates, and thus to understand the origin of life. To the contrary, I would argue that clarity of language is most important when we remain confused about our subject matter (Mix 2015; see also Millstein 2009). It is worth having clear life-concepts and, in the case of vegetable life, one is available.

Population Thinking

Regulator and replicator life-concepts attempt to start with an agential individual. Like the Platonic souls of the Middle Ages, they focus on a bounded individual with faculties or capabilities. Biological nominalist accounts begin with the interaction between population and environment. With Aristotle, they focus on an activity (in action and in fulfillment), specifically nutrition or reproduction, but see it in context. With Darwin, they remain agnostic as to the origins of the first vegetable population, displaying nutrition and reproduction. It is not enough for an organism to incorporate matter; it must incorporate matter in competition with other organisms. In a very important way, an organism alone is no organism.

The most natural kinds, from the perspective of evolutionary biology, occur on spatial and temporal scales that allow us to understand how populations change through time. They have persistent, identifiable units of inheritance. They vary in ways that impact their survival and reproduction. They occur in groups that compete for resources. In evolutionary biology, evolving populations are always etiologically prior to individuals.

Every organism exists as part of a lineage, running from the distant past. Its traits reflect millennia of interactions. At this point, the divergence from Aristotle becomes most clear. Aristotle thought that the form or essence of each species persisted in eternity, but Darwin suggested that populations continuously change. Ruth Millikan (2017, pp. 17–22) speaks of *historical kinds* with a common origin and causal dependence. A lineage, whether genetic, organismal, or species, holds together as a continuous chain of nutrition through time—punctuated, divided, and rejoined through reproductive events. Millikan treats these as "real kinds" suggesting more than pragmatic classification. Historical adaptation leads to "biological function" and more natural kinds (Millikan 1984).

This allows us to say the cells in a human body form a population. For timescales longer than the human lifetime, competition between them is mostly negligible because they all depend on a few germline cells for long-term survival. They all have the same evolutionary interests. This means that we are more likely to call human bodies organisms and individuals. And yet, we cannot rule out competition between cells when looking at cancer over shorter timescales.

We can identify populations of individuals that cooperate and compete for resources at a variety of timescales. Nutrition and reproduction blend together once we recognize that natural selection leads to any trait that, over time, increases the proportion of resources involved in its own propagation. On the shortest timescales, that means competitive nutrition. On longer timescales, it involves competitive reproduction as well. Some successful strategies lead to organisms with few offspring, but large bodies and long lifetimes. Other successful strategies lead to organisms with many, small, short-lived children. Both maximize the proportion of the environment caught up in the process. And, of course, strategies change as the environment changes. "There is variation and evolution of kind-hood, as well as of the organisms themselves" (Godfrey-Smith 2009, p. 15).

Peter Godfrey-Smith describes this as *evolutionary nominalism*: "the groupings of individuals into types is in no way essential to Darwinian explanation. Such grouping are convenient tools. But one always has the choice of using finer or coarser groupings, ignoring fewer or more differences between individuals" (ibid., p. 35). Millikan, similarly, speaks of different levels in biology as equally historical kinds. Addressing ant colonies and other "social individuals," she says this. "Each is something that one might keep track of and learn about over time" (Millikan 2017, p. 20). My biological nominalism may be more realist than Godfrey-Smith's and more pragmatic than Millikan's, but all three demonstrate a contemporary movement away from exclusive claims of individuality in biology. For an even broader range of arguments, see Millstein (2009) and Doolittle and Inkpen (2018) on populations of organisms as organisms themselves.

Non-exclusive Organisms

One organism may exist within another; one Darwinian population may exist within another. A population of genes competes within a bacterium. That bacterium is, itself, competing with other bacteria in a petri dish. Millikan and Godfrey-Smith speak of overlapping kinds. I see them as multiple, successful descriptions of a world that, for scientific purposes, lacks natural kinds.

Biological nominalism allows us to assign, for the purposes of any given experiment or theory, the labels of individual, organism, and population to replicators or regulators of any scale we wish. When we do so on the basis of an evolutionary causal nexus, for example, a gene spreading in a population of wolves, we have resurrected the vegetable soul. One process—evolution by natural selection—provides the efficient, formal, and final cause of the gene. As long as we pick a relevant timescale, the process is in action and in fulfillment. The fitness of traits to environment shapes lineages while the lineages shape both traits and environments.

Vegetable life requires neither individuals nor kinds in any essential sense. Evolution by natural selection allows us to account for organisms while maintaining the mechanical philosophy (as etiological reduction). Nominal forms suffice. These labels, nonetheless, represent a real process in the world. Once a population exists with regulators and replicators which cooperate and compete, natural selection will refine them, making them better at regulation and replication.

Regulators and replicators overlap in surprising ways. Both are organized. Both are more than physical aggregates; they evolve in ways that simple physical aggregates do not. Nutrition and reproduction differentiate them from their environment. And yet, in the wake of evolution, we must view these processes as occurring in and driven by the environment and a population, not just the individual. Organisms always exist in a particular context and, perhaps counterintuitively, changing the context can change the boundaries of the organism. Individuals cannot be separated out as agents or essences. Nor do they have natural limits that would allow us to say X cannot be an organism because Y is.

I want to be clear that I am speaking of organisms *from the context of modern biology*. Many will want to use the term 'organism' for ethical or theological purposes, or simply to articulate some concept of identity. I do not wish to rule out those uses. It is not clear to me, however, that these terms arise from biology. It remains possible that biology includes, within the broader category of life, a narrower category of exclusive individuals. We may find that we really do need a category of animal life that follows all the rules of vegetable organismality but also imposes an animal exclusivity. The next chapter will turn to individuality and subjective life-concepts. For now, it is worth noting that organismal or vegetable life forms a meaningful category in modern biology and modern science in general. Nutrition and reproduction, and hence evolution by natural selection, distinguish a clear set of phenomena, bigger than atoms and smaller than the universe. The concept of vegetable life is distinct, even when its application appears fuzzy. Concepts of vegetable individuality remain arbitrary.

REFERENCES

Benner, Steven A. "Defining Life." *Astrobiology* 10 (2010): 1021–1030.
Bernal, John D. *The Origin of Life*. Cleveland: World Publishing Company, 1967.
Cooper, Steven J. "From Claude Bernard to Walter Cannon. Emergence of the Concept of Homeostasis." *Appetite* 51 (2008): 419–427.
Dawkins, Richard. *The Selfish Gene*. New York: Oxford University Press, 1976.
Dennett, Daniel C. *Darwin's Dangerous Idea: Evolution and the Meanings of Life*. New York: Simon & Schuster, 1995.
DeWoody, Jennifer, Carol A. Rowe, Valerie D. Hipkins, and Karen E. Mock. "'Pando' Lives: Molecular Genetic Evidence of a Giant Aspen Clone in

Central Utah." *Western North American Naturalist* 68, no. 4 (2008): 493–497.

Doolittle, W. Ford, and S. Andrew Inkpen. "Processes and Patterns of Interaction as Units of Selection: An Introduction to ITSNTS Thinking." *Proceedings of the National Academy of Sciences* 115, no. 16 (2018): 4006–4014.

Dyson, Freeman J. *Origins of Life*. Cambridge: Cambridge University Press, 1985.

Godfrey-Smith, Peter. *Darwinian Populations and Natural Selection*. New York: Oxford University Press, 2009.

Hagedoorn, A.L. *Autokatalytical Substance: The Determinants for the Inheritable Characters, A Biochemical Theory of Inheritance and Evolution*. Leipzig: Verlag von Wilhelm, 1911.

Haig, David. "The Strategic Gene." *Biology and Philosophy* 27, no. 4 (2012): 461–479.

Haldane, J.B.S. "The Origin of Life." *The Rationalist Annual* 3 (1929): 3–10.

Johannesson, Hanna, and Jan Stenlid. "Nuclear Reassortment Between Vegetative Mycelia in Natural Populations of the Basidiomycete *Heterobasidion annosum*." *Fungal Genetics and Biology* 41 (2004): 563–570.

Joyce, Gerald F. "Foreword." In *Origins of Life: The Central Concepts*, edited by D.W. Deamer and G.R. Fleischaker. Boston: Jones and Bartlett, 1994.

Keller, Evelyn Fox. *Making Sense of Life: Explaining Biological Development with Models, Metaphors, and Machines*. Cambridge, MA: Harvard University Press, 2002.

Luisi, Pier Luigi. "About Various Definitions of Life." *Origins of Life and Evolution of the Biosphere* 28, no. 4–6 (1998): 613–622.

Mayr, Ernst. "Cause and Effect in Biology: Kinds of Causes, Predictability, and Teleology Are Viewed by a Practicing Biologist." *Science* 134 (1961): 1501–1506.

Mayr, Ernst. *The Growth of Biological Thought: Diversity, Evolution, and Inheritance*. Cambridge, MA: Harvard University Press, 1982.

Midgley, Mary. *Evolution as Religion*. New York: Routledge, 1985.

Miller, Elaine P. *The Vegetative Soul: From Philosophy of Nature to Subjectivity in the Feminine*. Albany, NY: State University of New York Press, 2002.

Millikan, Ruth G. *Language, Thought, and Other Biological Categories: New Foundations for Realism*. Cambridge, MA: MIT Press, 1984.

Millikan, Ruth G. *Beyond Concepts: Unicepts, Language, and Natural Information*. Oxford: Oxford University Press, 2017.

Millstein, Roberta L. "Populations as Individuals." *Biological Theory* 4, no. 3 (2009): 267–273.

Mix, Lucas J. "Proper Activity, Preference, and the Meaning of Life." *Philosophy and Theory in Biology* 6 (2014). http://dx.doi.org/10.3998/ptb.6959004.0006.001.

Mix, Lucas J. "Defending Definitions of Life." *Astrobiology* 15, no. 1 (2015): 15–19.

Monod, Jacques. *Chance and Necessity.* New York: Vintage Books, 1972.

Muller, H.J. "The Gene as the Basis of Life." *Proceedings of the International Congress of Plant Sciences* 1 (1926): 897–921.

Muller, H.J. "The Gene Material as the Initiator and the Organizing Basis of Life." *American Naturalist* 100 (1966): 493–517.

Ostrander, Elaine A., Brian W. Davis, and Gary K. Ostrander. "Transmissible Tumors: Breaking the Cancer Paradigm." *Trends in* Genetics 32, no. 1 (2016): 1–15.

Roper, Marcus, Chris Ellison, John W. Taylor, and N. Louise Glass. "Nuclear and Genome Dynamics in Multinucleate Ascomycete Fungi." *Current Biology* 21, no. 18 (2011): R786–R793.

Sagan, Lynn. "On the Origin of Mitsing Cells." *Journal of Theoretical Biology* 14 (1967): 225–274.

Schrödinger, Erwin. *What Is Life? The Physical Aspect of the Living Cell; with, Mind and Matter & Autobiographical Sketches.* Cambridge, UK: Cambridge University Press, 1992.

Shields, Christopher. "The Dialectic of Life." *Synthese* 185 (2012): 103–124.

Troland, L.T. "Biological Enigmas and Enzyme Action." *American Naturalist* 51 (1917): 321–350.

Watson, James D., and Francis H. Crick. "Molecular Structure of Nucleic Acids; A Structure for Deoxyribose Nucleic Acid." *Nature* 171 (1953): 737–738.

CHAPTER 19

Animal Individuality: The Subjective Self

Evolution by natural selection transformed biology into a modern natural science. A new life-concept, based on interactions between populations and environments, yields only arbitrary individuals. Their boundaries depend on which framework we choose to employ (in time, space, and population). Multiple options appear equally useful. In this way, the Aristotelian vegetable soul has been retained, with a caveat. Our formal causes have become historically contingent and non-exclusive. I describe this in the previous chapter as biological nominalism.

We now turn to Aristotelian animal souls and ask whether they can be redeemed in the same manner. Darwin sought to explain animals in terms of vegetable faculties and largely succeeded, but we must be careful in how we use the term 'animal.' The Kingdom Animalia can be described with biological nominalism, but the traditional category of animal life proves more difficult. Aristotle, and most subsequent authors, associated the animal soul with sensation and locomotion (as willed motion). I argue that these processes, and commonly associated subjective life-concepts, depend on a metaphysical interiority that cannot be provided by biological nominalism.

To be clear, I am arguing that modern biology will not accommodate the *traditional* faculties of sensation and locomotion. The lack of objective formal causes means that subjective interiority must be truly subjective. We can readily sidestep this issue by redefining sensation, will, and locomotion or by expanding our epistemology beyond biology as natural science. To the extent that sensation and locomotion have been

© The Author(s) 2018
L. J. Mix, *Life Concepts from Aristotle to Darwin*,
https://doi.org/10.1007/978-3-319-96047-0_19

redefined, those redefinitions should be made explicit. Projecting them onto biological discussions before Darwin will cause confusion. To the extent that we add to our epistemology, we must articulate what we have added and why.

After Darwin, many attempts have been made to define animal faculties in a less problematic way. Space does not allow a thorough treatment. Therefore, I must apologize for both brevity and limited expertise. Nonetheless, I believe readers will be interested in how biological nominalism and the vegetable soul relate to contentious issues such as the evolution and naturalness of consciousness. I have exercised my prerogative as an academic. At the end of a history of vegetable life-concepts, I speculate on broader implications: how they relate to animal life-concepts. I believe that Aristotle's vegetable souls can be revived, but that his animal souls cannot. I hope my remarks make my own perspective clearer.

PROCESSES WITH A BACK END

Traditional sensation and locomotion bridged embodied physical stuff with something less tangible. Confusion arises when we conflate the front end of the process, invariably embodied, with the back end, a more abstract soul. Within Platonic continua, the soul was *less* embodied. Even in Epicurean and Stoic circles, the soul was differently embodied, being made of rarified particles. With the rise of more dualist conceptions, the soul became subsistent and disembodied.

No one doubted vegetable embodiment or chains of efficient causation within vegetable bodies. In Aristotelian language, sensation involves external forms acting on bodily organs. Through the intellect, the mind perceives forms directly. Sensation differs from intellect because it perceives forms indirectly, via sensory organs in an animal body. Locomotion, or willed motion, similarly always included physical events like the contraction of muscles or the opening of flowers.

Animal life-concepts must invoke something more, though what that something was changed through time. In Platonic thinking, it was often associated with higher participation in cosmic life. From Augustine onward, this back end of sensation and locomotion became associated with subjectivity. It can be identified through self-reflection, but never by external observation. The back end defines what I refer to as metaphysical interiority and animal individuality. If it exists in some, but not all, organisms and can be detected or inferred empirically, then there will be some analogue of traditional animal life in modern biology.

All or Nothing

I object to animal and subjective life within modern biology, but not from a naïve objection to immaterial or unobservable processes. Rather, I worry about the extent of the category. If we take a reductive physicalist position and simply remove the back end (qualia and will), then our description of sensation and locomotion applies to all organisms. All living things respond mechanically to stimuli. All living things have a front end. Animal life becomes equivalent to vegetable life. If we retain the back end, but make it fully mechanical, the same problem applies, though one might arbitrarily define a particular kind of feedback mechanism. We do observe greater sensitivity and proto-agency in some organisms, but it has arisen in many different ways in many different lineages. We would need to account for sight in the Animalia, phototropism in Plantae, chemotaxis in protists, conjugation in bacteria, selective infection by viruses, and so forth ad nauseam.[1] We would end up with many different "animal" categories.

On the other hand, if we retain a traditional back end that processes stimuli immaterially (e.g., traditional intellect or will), it is unclear to me how we might determine this empirically. With Descartes, we might assign an immaterial soul to humans, but how would we rule out its presence in worms, bacteria, and viruses. The move provides little or no traction bridging human and vegetable life. Soulless animals again dissolve into vegetables while ensouled animals blend with humanity. Animals must either be too broad or too narrow to be of use.

ALTERNATE INTERIORITY

Throughout this history, the language of subjectivity has invoked interiority. Aristotle characterized sensitive life as being affected from without. Augustine spoke of "internal motivation" in animals, an inner self in

---

[1] Phototropism refers to responses to light stimuli. Chemotaxis refers to movement in response to a chemical gradient. Conjugation refers to a specific mechanism some bacteria have for exchanging nucleic acids with similar bacteria. Viruses recognize, attach to, and infect cells with specific coat proteins.

humans. Recent approaches in neuroscience and psychology also resort to the language of insides and outsides. In attempting to explain human decision-making in terms of brain function, Koechlin and Summerfield (2007, p. 229) define executive function or "executive control [as] the ability to select actions or thoughts in relation to internal goals." Baumeister (2008) refers to "inner processes." Even the most mechanical metaphors (i.e., the brain as a computer) invoke some sense of input and output.

It can be easy to conflate this subjective interiority with physical interiority, related to physical extension under the mechanical philosophy. A simple thought experiment demonstrates the difference. During brain surgery, the surgeon visually perceives the inside of the patient's brain but has no access to the patient's mind. Conversely, the patient, whether asleep or awake, has experience of her mental state, but cannot see the physical inside of her brain. This is not a new realization. It was built into Augustine's definition. David Chalmers sums it up succinctly as the hard problem of consciousness.

> It is undeniable that some organisms are subjects of experience. But the question of how it is that these systems are subjects of experience is perplexing. Why is it that when our cognitive systems engage in visual and auditory information-processing, we have visual or auditory experience: the quality of deep blue, the sensation of middle C? How can we explain why there is something it is like to entertain a mental image, or to experience an emotion? It is widely agreed that experience arises from a physical basis, but we have no good explanation of why and how it so arises. Why should physical processing give rise to a rich inner life at all? It seems objectively unreasonable that it should, and yet it does. (Chalmers 1995, p. 201)

Nor does subjective interiority align with vegetable (organismic) interiority. Organisms must be defined relative to timescale, population, and environment. We allow multiple, overlapping organisms. Moving from vegetable interiority to subjective interiority will require privileging some perspectives over others.

I raise this point to emphasize the differences between physical interiority, organismal interiority, and subjective interiority. I do not wish to draw any strong ontological, or even epistemological, conclusions from the difference. There *may* be successful ways to move from one to another. All three *may* be reducible to material and efficient causes. Or they might be ontologically distinct and unrelated. Personally, I find no current accounts convincing, but I both hope for and support research into reconciliation. In the meantime, it will be necessary to speak clearly about how we use the terms.

Traditional concepts of sensation, locomotion, and reason all include both front end and a back end processes. The front end process of sensation brings an environmental form (external organismally and subjectively) inside the body (internal organismally, but still external subjectively). The back end process takes the further step of subjectively internalizing the form. Similarly, for locomotion, a back end process holds choices and preferences, while a front end process executes the decisions made. Even in the case of reason, the common sense was often located in the physical, material front end of an organism, while intellect was reserved to the back end.

We may choose to deny that any back end exists. It is possible that sensation mechanically connects to locomotion with no truly subjective component. But, for many historical thinkers, it was precisely this back end that defined animal life and the animal soul. We do them an injustice to deny that this was the case or to simply redefine sensation, locomotion, and reason in ways that lack this component. To do so makes it impossible to understand the history of the concepts. More problematically, even for the most unrepentant physicalists, we run the risk of losing millennia of commentary on the explicitly physical *vegetable* components of sensation, locomotion, and reason.

We have not suddenly discovered the organismic limits of mental reality. We have only clarified long-understood vegetable aspects of life. The debate about whether, and how, those aspects are sufficient to explain everything about life remains. In any case, questions of nutrition, reproduction, and organismic unity can be answered independently of that question. With Bacon, Kant, and Darwin, we can pragmatically address the vegetable questions in biology while bracketing the animal questions. For me, this highlights the importance of maintaining a field of biology largely insulated from the animal dilemma, including issues of subjectivity and will.

EVOLUTIONARY REFINEMENTS IN INDIVIDUALITY

I have argued that biological individuals need not be exclusive. Perhaps they can evolve exclusivity. I can almost hear some of my readers reaching for specific theories of organism and how they lead to more discrete individuals. As long as they are transparent about the framework they use, I am delighted. I argued in the last chapter that some categories are more natural than others. Some evolutionary theories of individual and population provide better candidates for the interiority associated with the traditional animal life concept. Godfrey-Smith (2009, pp. 87–108) suggests *collective reproducers,* biological individuals composed of simpler individuals that reproduce as a group. Such collaboration allows for higher complexity and strategies to limit opportunistic nutrition and reproduction by the components. He speaks of such collectives "de-Darwinizing" component organisms by making individual persistence (at a given timescale) dependent on group persistence. This strengthens claims of interiority, unity, and (proto-)agency for the collective at the expense of other levels. We still must ask why subjectivity is observed (or attributed) at one scale and not another, but we can imagine a very complex form of organismal interiority as an intermediate.

Biological consortia look promising, but they can also defy common intuitions about individuals and species. Questions of biological agency have been complicated in the twenty-first century by the revelation that biological activities regularly occur only through the dynamic interaction of multiple species. Ed Yong (2016) describes many cases where biological activities require collaboration across species lines. What we think of as a human body requires more than just *Homo sapiens* cells and genes. Vast populations of bacteria, protists, and even tiny insects live on and in us. In many cases, organisms cannot live alone, at least not in any normal environment.

Obligate symbioses (species that cannot live apart) appear for nutrition, reproduction, sensation, and locomotion. Animals that feed on grass, like cows and goats, cannot digest the cellulose in plant walls. They require bacteria, protists, and fungi in their guts to turn it into useful energy and materials for growth. A more extreme nutritional symbiosis occurs between the citrus mealybug and two species of bacteria (p. 201). *Tremblaya* lives inside the mealybug cells, and *Moranella* live within the *Tremblaya* cells. Each genome alone lacks the genes necessary for nutrition. No one "organism" can produce all the enzymes it needs. The three collaborate, however, and function successfully. Aspects of nutrition, essential to each (both as ramet and genet) have been outsourced to other organisms.

More disturbing to traditional organism concepts, several cases of organismic hijacking have now been observed. Organisms can take over other organisms. The victims continue to regulate themselves, but only for the sake of another organism; they become zombies (Andersen et al. 2009). Spores of the fungus *Ophiocordyceps* land on ants. They use enzymes to dissolve a hole through the exoskeleton and grow into the body cavity. The mechanisms are still poorly understood, but the fungus does not immediately kill the ant. Instead, the ant climbs a plant and affixes itself to the bottom of a leaf, in a position congenial to the fungus. At this point, and only at this point, the fungus kills the ant and grows a fruiting body out of its back. The fungus directs ant sensation and locomotion for its own ends. Critically, the regular process of nutrition/metabolism must continue within the ant, If the whole body and metabolism of the ant serves the fungus, should it be counted as "inside" the fungal organism? How do we understand subjectivity and proto-agency in such a system?

Evolved Intention

Ruth Millikan and Daniel Dennett argue for intention and intentional systems arising through natural selection. They make clear, however, that these do not align with traditional animal life, specifically in terms of unity and interiority. Millikan (1984, p. 93) argues that "intentionality is grounded in external natural relations." Intention, like biological purpose, is only meaningful in the context of environment and population, a causal nexus of historical events. With Dennett (2009), she expressly denies the interiority and self-reflection invoked by Augustine and Descartes. This new intentionality departs from subjective interiority at precisely the point animal-life was meant to address. Nor am I convinced that their intentional systems map unambiguously onto more familiar replicators or regulators.[2] They may not even cluster into the unified

---

[2] In *Darwin's Dangerous Idea*, Dennett (1995) claims that a person is a new entity formed when memes infect an animal (p. 341). Later he refers to himself as both a person and animal (p. 426). A person must be either an animal alone or a consortium of animal and memes. One might claim that both brute (pre-meme *Homo*) and person (post-meme *Homo*) are animals, but without the reproductive bottlenecking that de-Darwinizes constituents, the memes appear to be properly classified as symbiotes and not components or organs of a larger animal.

centers of reflection and agency we identify with minds.[3] They may map onto biological consortia.

PANPSYCHISM

Some readers will associate vegetable souls with panpsychism. Panpsychists hold that mentality of some kind (e.g., reason or subjectivity) pervades the natural world (Goff et al. 2017). Minds or fundamentally mental mind-constituents can be found throughout the universe. For most panpsychists, this applies even at the most fundamental level. Thus, we either have minds all the way down (as with Leibniz' monads) or mind atoms (as with Lucretius' soul seeds). Many perspectives I have covered are amenable to such analysis.

Most panpsychists, historically, took a Platonic approach. They saw a single, transcendent soul informing all matter (e.g., Plotinus).[4] In the nineteenth century, panpsychists argued from idealism (e.g., Josiah Royce) or evolution (e.g., Haeckel), believing that a continuity of bodies suggested a continuity of minds. Notable examples include Heackel's *recapitulation*, Haacke's *orthogenesis*, and Chardin's *noogenesis*. Deductive arguments from panpsychism to biology appear to be off-limits a priori. Analogical arguments in both directions have not proved fruitful in·terms of predictions and, thus, have largely been rejected by biologists.

More recent panpsychists have been more inclined to speak of minds as psychic aggregates in parallel with physical bodies.[5] Here, Alfred North Whitehead (1861–1947) has been significant, building from events and processes, rather than substances. Such an approach may be more amenable to Aristotelian approaches and biological nominalism, but still represents a significant departure from mechanical conceptions of nature. Aspects overlap with the neutral monism of William James (1842–1910). Freya Mathews (2003) also adopts the label panpsychism in denying the Cartesian theater. Mathews argues for panphysicalism as well as panpsychism, creating a new hylomorphism, though one of matter and interiority in place of matter and form.

---

[3] It should be added that some psychologists now argue that unified consciousness is an illusion. See, for example, Gazzaniga (2012) and Kahneman (2011).

[4] Appeals to a cosmic or world-soul have been called synecological panpsychism and cosmopsychism.

[5] Appeals to soul constituents have been called atomistic panpsychism and micropsychism.

The traditional category of animal life comes to apply generally to vegetable systems, or equivocally to different branches of the tree of life. Within the context of modern biology, it answers neither Greek questions of individual persistence nor medieval questions of individual agency. The standard model of agency requires a discrete autonomous agent. Many practical questions in ethics, politics, and economics require us to identify exclusive individual subjects—moral patients, loci of rights and responsibilities, loci of preference. Recall that Locke differentiated between human bodies (brute animals) and persons (subjective agents). He was, no doubt, echoing the divide between mortal souls and immortal souls in Aquinas and the dual creation in Augustine. Having made intentions as well as bodies indiscrete and contextual, we can no longer map either onto discrete persons. The mechanical philosophy may preclude indubitable individuals.

I aim for a minimal point. Moving from organismic to subjective life represents a significant step, linguistically and epistemologically. Whether or not this reflects an ontological leap has been a subject of debate for thousands of years. Animals have always represented an awkward middle ground. Physiological connections between vegetables and animals are easy to draw. Psychological connections between animals and humans are similarly easy to draw. The hard problem, now as always, involves making both connections at the same time.

We face a difficult choice. We can accept continuity of organisms and persons, both of which have only arbitrary individuality. Or, we can preserve discrete persons but admit that we require something more than natural science to draw the boundaries. The first option unites physiology with psychology. The second reaffirms that they are distinct enterprises. Neither allows us to keep animal life as an intermediate between human persons and life at large. This may explain why the etymology of animal—as ensouled being—has been forgotten and why the vegetable soul sounds so alien to modern ears.

REFERENCES

Andersen, Sandra B., Sylvia Gerritsma, Kalsum M. Yusah, David Mayntz, Nigel L. Hywel-Jones, Johan Billen, Jacobus J. Boomsma, and David P. Hughes. "The Life of a Dead Ant: The Expression of an Adaptive Extended Phenotype." *The American Naturalist* 174, no. 3 (2009): 424–433.

Baumeister, Roy F. "Free Will in Scientific Psychology." *Perspectives on Psychological Science* 3 (2008): 14–19.

Chalmers, David J. (1995). "Facing Up to the Problem of Consciousness." *Journal of Consciousness Studies* 2, no. 3 (1995): 200–219.

Dennett, Daniel C. *Darwin's Dangerous Idea: Evolution and the Meanings of Life.* New York: Simon & Schuster, 1995.

Dennett, Daniel. "Intentional Systems Theory." *The Oxford Handbook of Philosophy of Mind*, 339–350. Oxford: Oxford University Press, 2009.

Gazzaniga, Michael. *Who's in Charge? Free Will and the Science of the Brain.* London: Hachette, UK, 2012.

Godfrey-Smith, Peter. *Darwinian Populations and Natural Selection.* New York: Oxford University Press, 2009.

Goff, Philip, William Seager, and Sean Allen-Hermanson. "Panpsychism." In *Stanford Encyclopedia of Philosophy*, Winter 2017 ed. Stanford University, 1997–. https://plato.stanford.edu/archives/win2017/entries/panpsychism/.

Kahneman, Daniel. *Thinking, Fast and Slow.* New York: Macmillan, 2011.

Koechlin, E., and C. Summerfield. "An Information Theoretical Approach to Prefrontal Executive Function." *Trends in Cognitive Sciences* 11, no. 6 (2007): 229–235.

Mathews, Freya. *For love of matter: A contemporary panpsychism.* Albany, NY: SUNY Press, 2003.

Millikan, Ruth G. *Language, Thought, and Other Biological Categories: New Foundations for Realism.* Cambridge, MA: MIT Press, 1984.

Yong, Ed. *I Contain Multitudes: The Microbes Within Us and a Grander View of Life.* New York: HarperCollins, 2016.

CHAPTER 20

What Can Be Revived (and What Cannot)

How do we account for the unusual organization and freedom of living things? Today's philosophers and biologists think of a theory or definition of life. For most of European history, it was a question of vegetable souls or vegetable natures. Nutritive souls, proposed by Aristotle, were the most popular theory. In the wake of the mechanical philosophy and natural science, these accounts lost their cogency, or perhaps we lost a language for making sense of them. In either case, new theories of life arose, centered on metabolism and evolution by natural selection.

Many have argued that we do not need a theory of life or that we must wait for one to emerge. I propose a more active approach (Mix 2015). As we look for life in space, investigate the origins of life on Earth, and explore the connections between all living things, we must learn to be clear about what we mean when we use the word, "life." The term cuts a broad swath in biological and social sciences as well as the arts and humanities. We can benefit across fields from a clearer articulation of how we think about life as well as the beginning, end, and quality of individual lives. Biologists, in particular, should either commit to a particular theory or identify the contingency and diversity of working principles.

Having said that, a clear articulation need not require discrete boundaries. Chemistry and physics have benefited immensely from probabilistic descriptions of events: unambiguous statements about ambiguous aspects of reality. Our understanding suffers when we attempt to shoehorn a messy reality into neat categories. It also suffers when we persist in using

© The Author(s) 2018
L. J. Mix, *Life Concepts from Aristotle to Darwin*,
https://doi.org/10.1007/978-3-319-96047-0_20

words and theories that hide our true thinking behind equivocation and generalities. Only by persisting in the difficult process of generating, clarifying, and testing theories of life will we come to useful knowledge.

The nutritive soul remains one of the best descriptions of organismality or vegetable life. Stripped of Platonic accretions and reinterpretations, the core concepts address our intuitions about organization and purpose. Darwin's theory of evolution by natural selection explains the complex organization of extant life. Kant's causal nexus unifies natural selection (as a string of efficient causes) and the apparent purposiveness of life (as a final cause). We need only make this process the identity or essence of life (as formal cause). Then we can say that a single process, in action and in fulfillment, acts as the efficient, formal, and final cause of living beings. Life is an active, chemical process whose cause, identity, and purpose is self-perpetuation.

I diverge from Aristotle on questions of identity and mechanism. We should not think of a subsistent entity perpetuating itself through nutrition. Instead, we should think of a self-perpetuating process of nutrition. Life is not a set of things that have nutrition; life is nutrition. We can adopt a biological nominalism, in which the boundaries between individuals change depending on circumstances and, critically, the interests of observers.

Animal and rational souls do not fare as well, at least within the context of scientific thinking. Neither do the subjective and spiritual life-concepts developed over the centuries. This is not to say that they are bad ideas. They may be terribly useful philosophically. They are likely indispensable theologically. They are not, however, amenable to mechanical and empirical analysis. They largely depend on exclusive ideas of interiority that invoke both formal and final causes in precisely the manner rejected by the precursors of modern science. Limiting biology *as natural science* to the vegetable life-concept, we can focus the attention of biologists on empirically tractable problems, while freeing philosophers and theologians to address other life-concepts using other means.

LIFE AND DUALISM

Vegetable souls can help us understand the place of humans in the context of living things and in the context of a physical universe. We continue to ask the same questions proposed by the ancient Greeks, questions about composition, agency, individuality, and purpose. There have always been attempts to unify life. Souls provided the link for most of Western history, though their interpretation has changed repeatedly. We can trace the development of five distinct life-concepts: vegetable, animal, rational, subjective, and spiritual. And we can see how concerns about rational, subjective, and spiritual life led many to separate them from vegetables and physical "nature." Animal life often hung problematically in between. None of this should distract us from the rich discussion of fully natural, mortal processes associated with vegetable souls.

Over the centuries, two distinct languages developed, one physiological, the other psychological; one linking physics, plants, and animals and another linking animals, humans, and different concepts of the divine. Two approaches arose to deal with the gap. Platonic approaches tried to bridge the gap by making vegetables more rational, subjective, and spiritual. Aristotelian approaches physicalized animals, including humans.

Descartes sundered the world with a mechanical philosophy that intentionally kicked souls out of the physical realm, taking true agency, individuality, and purpose with them. The move represented a radical departure and was never fully accepted. Philosophers immediately tried to reframe it in terms that allowed for organized matter, embodied minds, and causal links. Kant and Darwin provided great insights into a mechanical biology. Starting with an etiological reduction that sidestepped formal and final causes, they proved the effectiveness of a pragmatic approach.

The Composition of Life

Questions our history of life-concepts began with classical questions of change and persistence. Living things persist through change. One might argue that an organism can even persist through a complete replacement of atoms over the course of its lifetime. We need a concept for whatever-it-is that provides continuity, a vegetable soul. Vegetable soul concepts, almost universally, claimed that vegetable life can only be encountered in the context of physical matter. In many cases, the vegetable soul was both material and physical, made of the same stuff and formed into the same kind of body, as non-living things. In other cases, the vegetable soul was an extension of a cosmic soul working out its life through all matter, not just living things. Vegetable souls should not be considered immaterial, unnatural, or subsistent. With rare exceptions, they have always been intended to be consistent with the rules that apply to the non-living world.

Physics applies. We must ask, then, whether physics is sufficient. This returns us to Mill's question about nature. If we define nature to exclude organisms, they shall be excluded. If we define nature to include them, they shall be included. Platonic and Neoplatonic approaches suggest that no explanation will be sufficient without invoking a transcendent agency that organizes and directs life. Such agency seems unnecessary, given evolutionary redefinitions of formal and final causes. It may exist, but biology and vegetable life can get by without it. Aristotelian approaches, focused on physical properties and processes look more promising.

The same issue arises for animal life. Almost universally, scholars across fields said that vegetable rules applied to animals. We must ask whether they are sufficient. What do we want from the animal category? The mechanical philosophers intentionally excluded rationality, subjectivity, and spirituality from "nature." To put them back in will require renegotiating the boundaries of natural science.

Natural Selection and Efficient Causes

Darwin's theory spoke of the *means* by which living things, their organization, and function, come to be. Darwin, himself, remained self-consciously silent about the motivation behind the means. Speciation through natural selection provides proximal efficient cause accounts for biological activity. Starting with the most basic Darwinian population, we can imagine the evolution of highly sophisticated biological mechanisms. These include extremely refined and efficient nutrition and reproduction as well as a front end for sensation, locomotion, and reason.

Following the lead of Buffon and others, Darwin spoke of the dynamic interaction of populations with an environment. Departing from them, he avoided giving agency to either. For Darwin, every living thing had a context, a place and time within a population and an environment. Every living thing sprang from parents and interacts with other living things, cooperating and competing to use limited resources. If a trait helped an organism to survive and reproduce, then it would outperform other members of the population. If it can pass on that trait, then its offspring will expand in the group through the generations. Population-level evolution, rather than individual agency, drives biology.

CAUSAL NEXUS AND FINAL CAUSES

Darwin embraced the causal nexus of Kant. Once a population of vege-table proto-agents exists, it will develop greater and great organization and purpose. The process refines vegetable life, making individuals better and better at nutrition and reproduction. Present life, which appears so beautifully designed and elegantly focused, can arise from the most basic population. We cannot eliminate the ultimate cause problem, but we can make it ever so much simpler.

Modern evolutionary theories include final causes but do so as a gloss on the chain of efficient causes that make up natural selection. In Aristotle's language, final causes act through efficient causes. In Kant's language, the two reciprocally reproduce one another. Both Aristotle and Kant likely held some notion of a subsistent entity motivating the process, an essential form or noumenon bringing it about. These ideas were part of their intellectual environment. But, their theories do not require it, and the pragmatism of Bacon and Darwin can do without. A final cause requires no agential self, will, or intention. It needs only a "that for the sake of which." In evolution by natural selection, the final cause is "for the sake of persistence" within a particular timeframe, popu-lation, and environment. A trait is adapted when it has proven successful at this. Its biological purpose is ever and only to continue succeeding at this. Living things have such elegant, expansive, and surprising freedom because they have grown exceptionally good at nutrition and reproduc-tion. They occur, nonetheless, within a cycle of efficient causes.

Causal Nexus in Modern Biology

I am not the first to propose this causal nexus linking final causes to a string of efficient causes. I see this explanation as consistent with the excellent work of others, namely Ernst Mayr and David Haig. My inter-est, here, is to place the theory in historical context as the rightful heir of the vegetable soul. It remains consistent with the thoughts of Aristotle. Nor is this entirely surprising.

Mayr's famous paper on "Cause and Effect in Biology" creates a causal nexus with *evolutionary programs*, brought about by historical events. He suggested a reciprocal relationship between the formation of programs through a string of historical efficient causes and the purpo-sive execution of the programs (Mayr 1961; Mix 2016). While he found final causes associated with both in Aristotle, he thought only the latter (which he calls *teleonomy*) could be called goal-directed. Mayr's use of "ultimate cause" and, elsewhere, "unmoved mover," suggests both Aristotelian inspiration and a traditional reification of the biological agent (Mayr 1974).

David Haig presents a more nuanced take on environment and probability, comparing biological information to Aristotelian forms and biological function to Aristotelian final causes.

> Darwinian final causes are similarly grounded in efficient causes and are perfectly adequate, indeed indispensable, for certain kinds of biological explanation. A 'selection pressure' summarizes many reproductive outcomes just as the pressure of a gas summarizes many molecular motions. Darwinism, like thermodynamics, is a statistical theory that does not keep track of every detail. (Haig 2014, p. 694)

Final causes describe the causal nexus, through which a statistical aggregate of diverse efficient causes, distributed in space and time, can be summarized. While nothing beyond the efficient causes has been added, the final cause language provides the appropriate level of resolution for satisfying accounts of organization and purpose. Haig also mentions the agency of genes but highlights their dual role as agents and patients in natural selection (Haig 2012).

By equating the final cause of life with a historical string of efficient causes, we can stay safely within the etiological reduction of the mechanical philosophy without losing a meaningful idea of biological function and activity. I must emphasize that I am not denying any further purpose or agency to life. I am only stating, with Bacon and Darwin, that this causal nexus is enough to be getting on with for an empirical biology. Given an initial population, an arbitrarily high level of organization, purpose, and proto-agency may be achieved with sufficient time and physical inputs.

PRAGMATIC FORMAL CAUSES

By distributing agency across the population and the environment, modern biology can do without subsistent souls and discrete individuals. Dispersed processes can be described in multiple ways, each of which reveals genuine evolution by natural selection. We can and should continue to argue about which descriptions are most useful (e.g., levels of selection, units of inheritance, replicators versus regulators), but it has become clear that disparate models can provide accurate description and useful prediction. Biological nominalism suffices. We have no reliable access to labels in the mind of God (or natural kinds in the universe). Thus, we are better served by basing our biological labels on the theory of natural selection than attempting the converse.

CATEGORIES, NOT CAUSES

I have attempted, throughout, to speak of descriptive categories rather than historical causes. I would say that we observe life in the context of matter and consciousness in the context of life. I am even comfortable *in the language of modern biology* saying that life and consciousness "arise" in that context, with the understanding that the dependence is descriptive and explanatory (etiological) and not essential (ontological). The language of physical, vegetable, animal, and rational remains popular because it captures something about the way we view the world. We still refer to inactive or unresponsive humans as vegetables; they still exhibit nutrition, but lack locomotion and sensation. We still refer to irrational or unreflective people as animals or brutes; they lack reason. The categories provide a language for critical thinking which may reveal crucial flaws in the categories themselves. Relationships between them should not be taken for granted.

Most Western thinkers historically, and most people in my experience, view living bodies as a subset of physical aggregates and associate subjective things with a subset of living bodies. Hence, the *scala naturae*, whether viewed as eternal emanation, historical unfolding, or the result of evolutionary diversification. We tend to think of life as a series of nested categories: physical, vegetable, animal, and rational. We also tend to assign them increasing value. I want to reveal the traditional hierarchy, not defend it. Modern research suggests that the categories should not be nested and, if I have done my job, this book provides historical examples to help modern researchers.

Rational, subjective, and spiritual life may occur outside of conventional biology. Recent work on artificial intelligence and memes challenges us to think about the meaning of these concepts beyond traditional bounds. Historical reflection on angels, demons, stars, planets, and gods can provide key insights into how we can and should think about "life" beyond the vegetable context.

Modern biology, biology as *natural* science, rests on a commitment to biology being a subset of nature and nature being knowable in a particular way—namely nominalism, mechanism, and empiricism. That commitment can be renegotiated (e.g., Plantinga 2011; Nagel 2012), but we must be clear about the extent of such an enterprise. We can draw on four centuries of attempts to integrate formal and final causes into nature and a longer history that predates "nature" as we now think of it.

My own preference, as a modern biologist, would be to preserve the vegetable life category and redefine it in terms of natural selection, but recognize the epistemic boundaries that make that move possible. Following Bacon, Kant, Locke, and Darwin, natural science should methodologically reject formal and final causes while remaining agnostic ontologically. Standard agency and subjective interiority remain off-limits a priori, and therefore cannot be either proven or disproven by biology *as natural science*.[1] Removing the epistemic firewall risks threatening the tremendous utility already discovered. In this register, the difference between non-living and living becomes clear: living systems can be described in terms of evolution by natural selection. Individual lives remain arbitrary designations, critically dependent on the environmental and temporal scope we choose.

Many will want more definite individuals. Indeed, I think they will be necessary for medicine, politics, ethics, and theology. I only ague that they cannot be founded on vegetable or biological life-concepts in line with Darwin. To ask biologists to provide them threatens biology. Other philosophical foundations must be provided and, because so little consensus exists, must be articulated clearly.

Nutrition and reproduction continue to define "biology" as both a subject matter and an epistemology. Biologists study regulation and replication and the organisms that exhibit them. Biologists use the process of natural selection to provide compelling accounts for why these entities

---

[1] "[T]he mechanistic picture of the world inaugurated by Galileo, Descartes and Newton put the problem of the mind at center stage while paradoxically sweeping it under the rug." Goff et al. (2017); see also Foucault (1994) and "Mind and Matter" in Schrödinger (1992).

have the properties they do. We now have a much clearer grasp on how populations evolve and how that evolution affects our choices.

We still have many questions. We do not currently know how the first evolutionary populations came to be. It *may* be necessary to invoke transcendent agency for this transition; Wallace did. And yet, it remains unclear how such an agency might be demonstrated or prove useful. Hume's critiques (and their many descendants) successfully refute intelligent design theories. For my part, I will put time and energy into exploring the origins question with physical science and an Aristotelian approach; both have proven successful in other areas. Perhaps we should invoke a *hard problem of organismality* in parallel to Chalmer's hard problem of consciousness. Crossing these divides will require more than acquiring additional data or coming up with new, creative hypotheses. It will require a clearer understanding of the way we use life-concepts, the purposes we want them to serve, and the history of the discussion.

REFERENCES

Foucault, Michel. *The Order of Things: An Archaeology of the Human Sciences.* New York: Vintage, 1994.

Goff, Philip, William Seager, and Sean Allen-Hermanson. "Panpsychism." In *Stanford Encyclopedia of Philosophy*, Winter 2017 ed. Stanford University, 1997–. https://plato.stanford.edu/archives/win2017/entries/panpsychism/.

Haig, David. "The Strategic Gene." *Biology and Philosophy* 27, no. 4 (2012): 461–479.

Haig, David. "Fighting the Good Cause: Meaning, Purpose, Difference, and Choice." *Biology and Philosophy* 29, no. 5 (2014): 675–697.

Mayr, Ernst. "Cause and Effect in Biology: Kinds of Causes, Predictability, and Teleology Are Viewed by a Practicing Biologist." *Science* 134 (1961): 1501–1506.

Mayr, Ernst. "Teleological and Teleonomic, A New Analysis." *Boston Studies in the Philosophy of Science* 14 (1974): 91–117.

Mix, Lucas J. "Defending Definitions of Life." *Astrobiology* 15, no. 1 (2015): 15–19.

Mix, Lucas J. "Nested Explanation in Aristotle and Mayr." *Synthese* 193, no. 6 (2016): 1817–1832.

Nagel, Thomas. *Mind and Cosmos: Why the Materialist Neo-Darwinian Conception of Nature Is Almost Certainly False.* New York: Oxford University Press, 2012.

Plantinga, Alvin. *Where the Conflict Really Lies: Science, Religion, and Naturalism.* New York: Oxford University Press, 2011.

Schrödinger, Erwin. *What Is Life? The Physical Aspect of the Living Cell; With, Mind and Matter & Autobiographical Sketches.* Cambridge, UK: Cambridge University Press, 1992.

Author Index

Subject Index